Making Shoji

Making Shoji

Toshio Odate

Photographs by Laure Olender

LINDEN PUBLISHING, INC.
FRESNO

Making Shoji

by

Toshio Odate

Photographs by Laure Olender
Front cover photo by John Kelsey
Drawings by Toshio Odate

ISBN 0-941936-47-3

PRINTED IN THE UNITED STATES OF AMERICA

3 5 7 9 8 6 4 2

Library of Congress Cataloging-in-Publication data

Odate, Toshio
Making shoji / by Toshio Odate
p. cm.
ISBN 0-941936-47-3
1. Shoji screens -- Design and construction. I. Title.

TH2544.O33 2000
749'.3--dc21 99-049812

LINDEN PUBLISHING

The Woodworker's Library

Linden Publishing Inc.
2006 S. Mary St.
Fresno, CA 93721
tel 800-345-4447
www.lindenpub.com

I DEDICATE THIS BOOK TO

MY MASTER YASUTARO SAKAMAKI

TO WHOM I DID NOT HAVE A CHANCE TO SAY

THANK YOU

Table of Contents

Publisher's Note

This book offers insight and understanding to those who would appreciate a non-Western cultural perspective as expressed through traditional craft. The author, Toshio Odate, was born in Japan in 1930 and during the immediate post-war years served an apprenticeship as a *tategu-shi*, a maker of sliding doors. He went on to study art and design at university, and in 1958 was awarded a fellowship for the introduction of traditional Japanese woodworking in the United States, where he has remained.

Over the following four decades, Odate taught industrial design and sculpture at the university level, exhibited his own sculptures, and demonstrated Japanese woodworking techniques and attitudes at many workshops, classes, and seminars. In 1984, to introduce traditional Japanese woodworking philosophy and attitudes to Westerners, he published *Japanese Woodworking Tools: Their Tradition, Spirit and Use* (also available from Linden Publishing). Because of his unusual life, Odate represents a rare bridge from the ways of a traditional apprenticeship in the Japanese countryside, to mechanized woodworking in the United States today.

The publisher and editors have been acquainted with Odate for more than 20 years, and have gained a deep respect for who he is and how he presents his ideas. Therefore, we have tampered as little as possible with the author's organization and presentation of his material. Toshio Odate is an authentic voice, best heard unfiltered.

Acknowledgments

I thank many of my friends who gave me their encouragement in the writing of this book. I especially appreciate the strong support of my publisher, and I thank the committee of the Pratt Institute Faculty Development Fund for their support.

I would particularly like to thank my assistant, Laure Olender, who gave great effort, patience, and enthusiasm to the difficult task of editing and organizing the text of this book, and who also took photographs with deep creative sensibility.

Toshio Odate

The *shoji* is ready for rice paper.

Introduction

Traditional Japanese wood structures, temples, shrines, houses, and gardens have all been well introduced to the Western world through books, articles, and photographs. As we know today, traditional Japanese houses do not have as much wall space as Western houses. Most of the rooms, as well as areas in between the exterior and interior of the house, are divided by doors and screens. Primary among these screens is the *shoji*.

In 1982 I wrote an article about making *shoji* in *Fine Woodworking* magazine. Because I had included so much information, there was much that could not be contained in a 9-page article. One important unpublished portion was the emphasis on the Japanese tradition of woodworking, especially in the interior of the house. This book will not only concentrate on how to build Japanese *shoji*, it will also explain Japanese tradition and its value.

Every country has its own traditions and each is meaningful and beautiful in its own way. Especially in woodworking, there are often very subtle places of beauty and rules to follow. If you ignore them, the resulting project can appear shallow and heartless in quality. In my own country, Japan, I have often seen Western culture being adopted very casually, so it loses its original meaning and value. Of course, when you try to bring another country's traditions or culture to your own, they will often need some modification or evolution. But if you understand the quality of the origin, then it is possible to create richness by adopting other cultural traditions.

Japanese Tradition & Craftsman's Attitude

Japanese woodworkers believe in, treat, and appreciate wood as a living material. This attitude extends to matching the use with the wood's original state. I read an essay written by the great temple carpenter Tsunekazu Nishioka. When he was repairing Horyuji Temple (the oldest known wooden temple in the world) he discovered that its main columns were standing in the same fashion as they were in the woods. In other words, the south side of the tree was the south side of the column. The ancient Japanese carpenter believed that since the south side of the tree faced the sun, the south side of the column should also face the sun. As a result of this theory, the column would last a longer period of time. The carpenter also used a tree over a thousand years old. When the temple was built it was made not to last just three or four hundred years but at least a thousand years. Today we cannot afford to build like this, however the attitudes and the manners have carried on until this day. I will write some examples here.

I have often seen in America craftsmen book-matching grain patterns on paneling or tabletops. This is not commonly seen in Japan—I only saw it once or twice. When you do book-matching, the one panel has to be reversed with the grain pointing outward, while the other panel's grain points in the opposite direction. For the Japanese, this is very unnatural.

Traditional Japanese craftsmen don't use sandpaper, although they had a material like sandpaper called *tokusa* (leaf) and also shark skin. These materials were used for similar purposes as sandpaper, but in very limited situations. Also, the Japanese craftsman doesn't use paint or any other kind of finish on most wooden pieces. However, the Japanese do have a paint called *urushi,* or Japanese-lacquer, which, again, is used only in limited interior places.

Japanese woodworkers mainly use so-called softwoods. Commonly they use cedar, pine, and *hinoki*-cypress. Japanese believe *hinoki*-cypress is the best material for the house and its interior sliding doors and screens, or *shoji*. People dreamed of building their houses completely in *hinoki*-

cypress, but only few could afford such splendor. These woods have a distinctive coloration between the heart, or core, and the sapwood. Of course, some hardwoods like cherry also have such coloration. When the craftsman uses wood for a frame or panel, he uses the core for the inside and the sapwood for the outside. For instance, as he makes the frame of a door, the cores of the rails and stiles are directed inward and the larger portion of the core, which indicates the bottom of the tree, is used on the bottom of the stile. And on all rails, the wide portion of the core should be placed on the same side, if possible, as shown in the drawing on this page (D1).

Japanese appreciate tight *masame*, or quarter-sawn grain, more than Westerners do, probably because, as I earlier mentioned, they don't use paint. Not only craftsmen knew this, but people in the common society shared this value too. Also, Japanese craftsmen do not reverse house columns or place any other standing material upside down.

Most of what I wrote so far is about the visual traditional aesthetic and its value. Next I will talk about a very important and delicate aspect of traditional interior work, mortise-and-tenon joinery. Japanese try not to show end grain in interior work. We believe the end grain of wood is neither delicate nor elegant, thus the interior sliding doors and screens strictly have blind mortises. Exterior sliding doors like *ama-do* or rain door, *garasu-do* or glass door, *genkan-do* or entrance door, do use through mortises. Despite using blind or through mortises, we don't use pins nor wedges, we try to minimize the visual mechanism of joints. Many craftsmen I know believe that the mechanics of exposed joints spoil the aesthetic value of interior woodworking. What I have mentioned here is a small detail of the Japanese woodworking tradition, but it is an important key to understanding Japanese aesthetic values and Japanese attitudes.

While holding these matters in mind, I will explain to you how to build a traditional Japanese *shoji*.

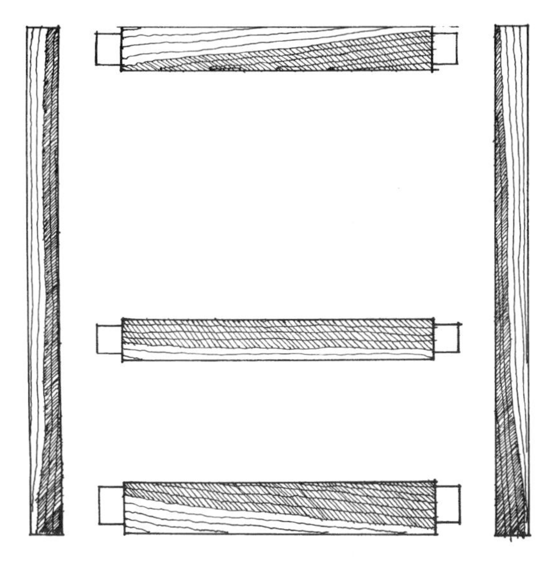

D1 The craftsman places the core, or heart, toward the inside and the sapwood toward the outside of the frame.

A Brief History of the Shoji

Before you start making a *shoji*, you should know some basic information about it. We don't know when and where the *shoji* came from, but we do know that early in history Japan had wooden exterior doors that had to be lifted up to be opened. Immediately behind these doors was another door, which had wooden lattice covered with paper for letting in light. I don't know whether the second door was lifted, was sliding, or had another mechanism. Some scholars say sliding doors derived from the exterior lifting wooden doors, others say they developed from interior room dividers.

We know that room dividers existed before sliding doors and panels. One type, called *byobu*, was made from free-standing folding panels covered with opaque paper. Another type, called *kicho*, was a free-standing frame covered with thin material usually placed in the middle of the room. Today *kicho* has turned into *tsuitate*, or free-standing room dividers, but *byobu* still exists as it did then.

By the 11th century, *shoji* had evolved into its present style. All sliding paper panels were then called *shoji*. From this general term, the translucent paper panel came to be called *akari-shoji*, or lightning *shoji*. Later on, the *akari-shoji* was just called *shoji* and the opaque paper panel was renamed *fusuma*.

Today we have many types of *shoji*. In the late 19th century, when glass was introduced to Japan, paper was set aside. But the glass *shoji* were too heavy, the soft wooden tracks couldn't take the weight. Since then, however, small portions of glass have been incorporated into *shoji*, and as a result several new and interesting styles have been created.

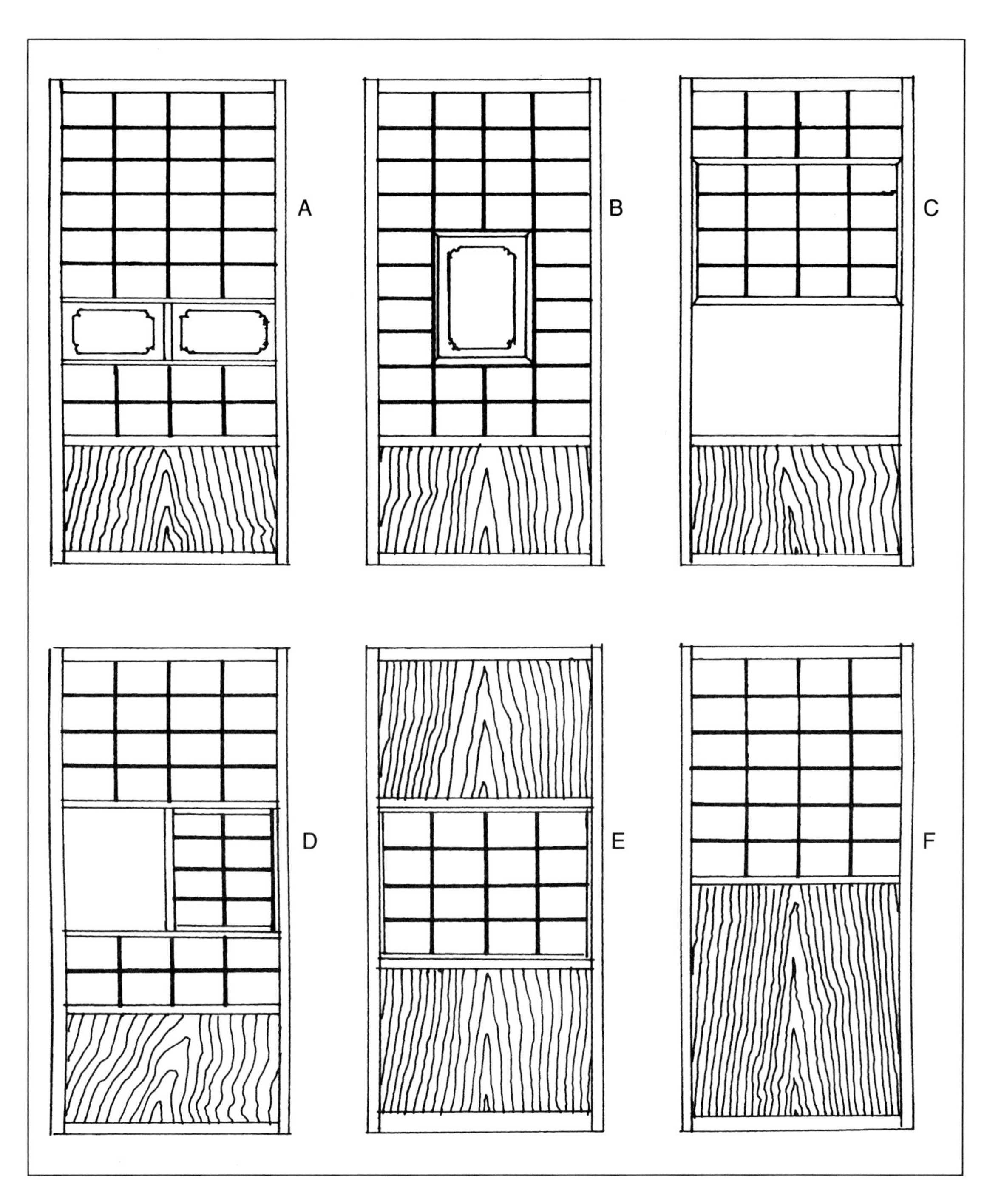

D2 Varieties of *shoji*.

A. Horizontal glass *shoji* (*yokogarasu-shoji*)

B. Picture frame *shoji* (*gakubuchi-shoji*) with glass panel

C. Snow-watching *shoji* (*yukimi-shoji*) with glass panel covered by vertically sliding panel

D. Cat-peeking *shoji* (*nekoma-shoji*) with glass panel covered by horizontally sliding panel

E. Middle *shoji* (*naka-shoji*) for dividing rooms

F. Entrance *shoji* (*koshidaka-shoji*), currently out of fashion

D3 Parts of the common *shoji.*

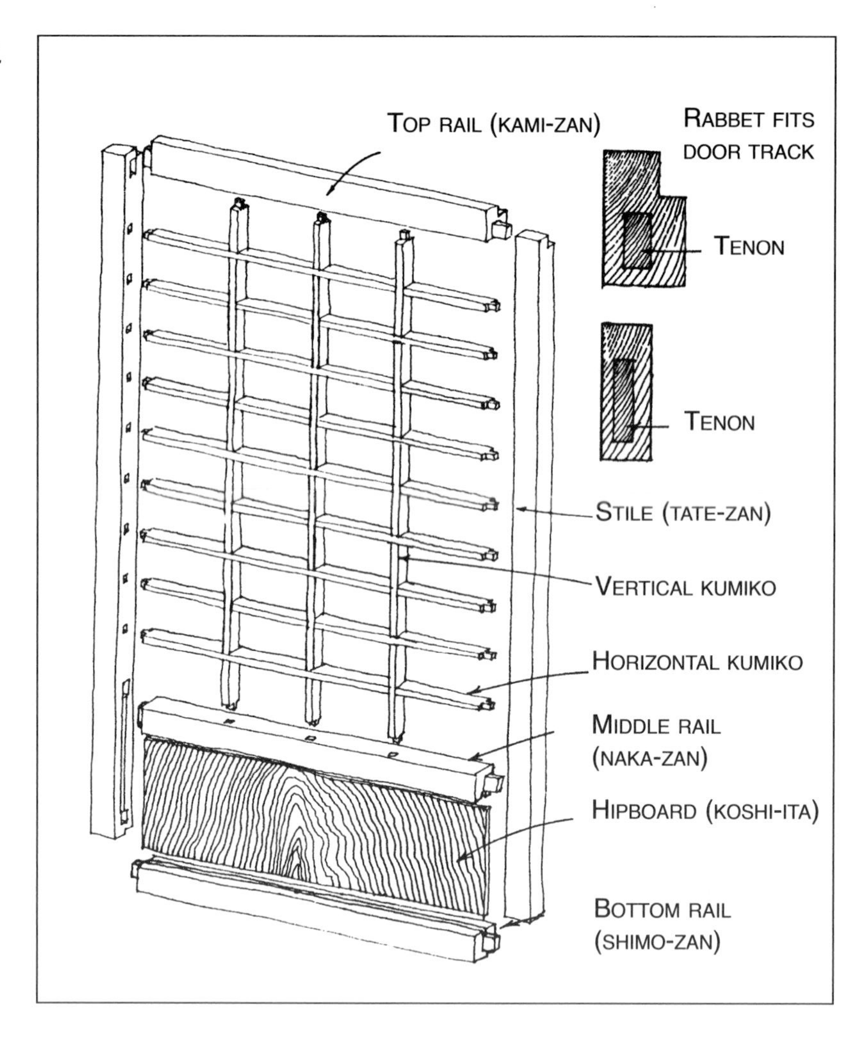

The Common Shoji

In Japan there are many types of *shoji.* The one best known to the West is the *ma-shoji,* or common *shoji.* Some Japanese also call this type of *shoji* blind *shoji,* because only translucent paper is placed on the wood.

The common *shoji* has two vertical stiles (left and right), three rails (top, middle, and bottom) ,and a hipboard. Inside the frame are the *kumiko,* or lattice, to which the *shoji* paper is applied (D3). Traditionally, there are two sizes of common *shoji.* One uses *mino*-sized paper (11″) (279mm) and the other *hanshi*-sized paper (9 3/4″) (248mm).

Mino paper *shoji* need three vertical *kumiko* pieces, nine horizontal pieces, the hipboard, and five strips of *mino* paper. *Hanshi* paper *shoji* need three vertical *kumiko* pieces, eleven horizontal pieces, a hipboard, and six strips of *hanshi* paper. If there are more than three vertical pieces and fewer than nine or eleven horizontal pieces, the *shoji* is called *tateshige shoji.* If there are three or fewer vertical pieces and more than nine or eleven horizontal pieces, it is called *yokoshige shoji.*

Shoji measure about 5′8″ (1730mm) high and 2′10″ (860mm) wide. I will tell you

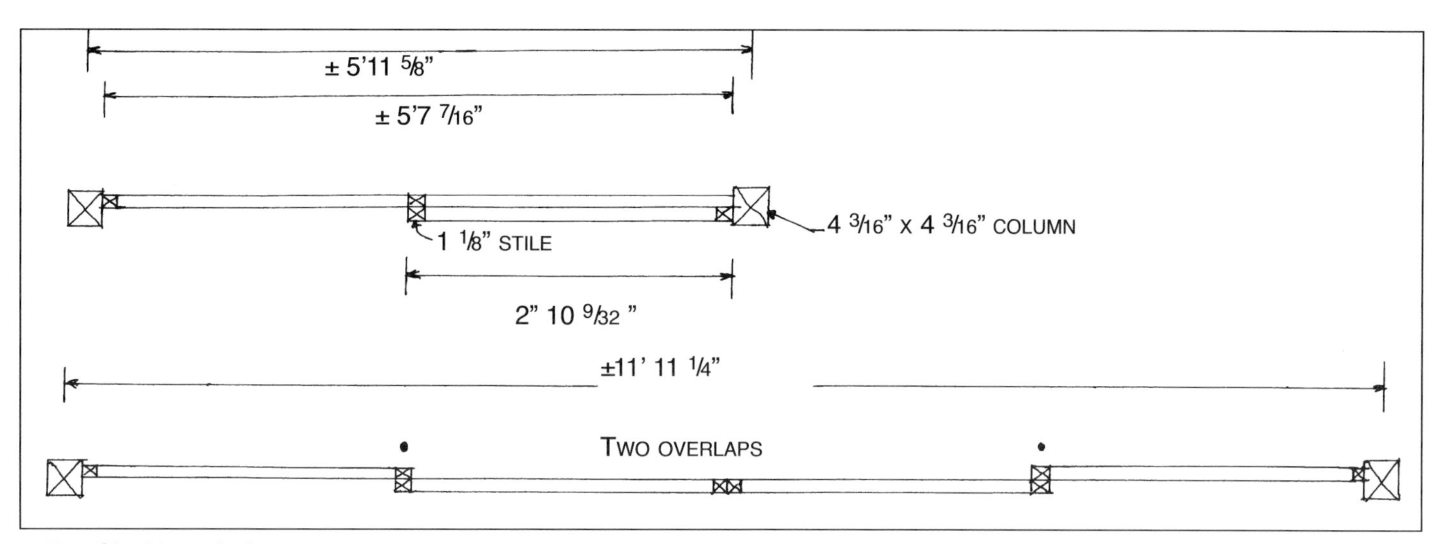

D4 *Shoji* installation.

how these proportions came about. Japanese interior openings depend upon the distance between the columns that support the roof. In height, most openings are five *shaku* and seven *sun* high, which is about 5'7 7/16" (1710mm). The width, measuring from the center of one column to the center of the next, can be 3, 6, 9, or 12 *shaku*. To find the width between columns, let us say that an opening from one column center to the next column center is 6 *shaku* (+- 5'11 5/8") (1820mm) and the thickness of the column is three *sun* and five *bu* (about 4 3/16" (106mm) square). Thus the opening will be 5'11 5/8" - 4 3/16" = 5'7 7/16" (1710mm). If your *shoji* stile's face is 1 1/8" (29mm) and you are making two *shoji* in between the 5'7 7/16" (1710mm) opening, you also have to add one stile width for the overlapping *shoji* (D4): 5'7 7/16" + 1 1/8" = 5'8 9/16" (1740mm). By dividing this sum into two, you will find the width of each screen—2'10 9/32" (870mm). To the height you also must add 5/8" (16mm) for the top track's groove. Thus the height of the screen would be 5'7 7/16 + 5/8"= 5'8 1/16" (1730mm).

The common measurements for the stiles and rails are:

	Face	Thickness
stile	1 1/8"	1 3/16"
top rail	1 7/16"	1 1/8"
middle rail	1 3/8"	1 1/8"
bottom rail	1 3/8"	1 1/8"

Thus the screen will measure 5'8 1/16" (1730mm) in height and 2'10 9/32" (870mm) in width. These measurements are for the proportions and look of a common *shoji*. Of course, in a Western home, you will need to adapt the dimensions to the openings you wish to fill. But you must follow this same logic and remember to account not only for the overlap in width but also for the height of the track at the top.

1-2 As you can see in these photos, *shoji* usually are made in pairs. The doors slide past one another, each in its own track. Thus it is necessary, when calculating the width of each *shoji*, to account for the overlapping stiles.

3 Laying out frame parts by snapping an ink line with the *sumitsubo*.

4 The *sumitsubo* is used in much the same way as the Western chalk line.

5 Sawing frame parts. The plank is propped up on one low horse and I am standing on it to hold it in place.

6 Planing the stile on the planing beam. Japanese planes cut on the pull stroke.

7-10 Slicing the hipboard in two with a handsaw. Board measures 18" by 30" by 7/8" (460mm x 760mm x 22mm).

PREPARATION

Traditionally, Japanese craftsmen use boards that measure 1 7/16" (37mm) thick, called *itawari*, to make the frames, and 7/8" (22mm) thick boards to prepare the 3/4" (19mm) thick *kumiko*. Because the *kumiko* are so small, *kumiko* boards must have straight grain and be of top quality. Unlike the *kumiko* boards, one has many more choices when choosing *itawari* boards.

It is a common saying among Japanese craftsmen that when an apprentice can accurately prepare materials for doors or screens, he is then very close to finishing his apprenticeship. People usually look at the finished product and say, "The craftsman is neat," or "He has skill," but actually most of the quality of the work is based on the preparation of materials.

Good preparation starts with the layout of the pieces on the wood. Check if the wood has any defects, such as knots, reaction wood, rot, or cracks. Cracks can derive from the separation of grain, worm holes, and even the rounded edge of the board. You need experience and knowledge to dissect the wood correctly, so you will know which part to use, which part could work, and which part is not to be used.

After you lay out the measurements on the wood, carefully start to saw out the pieces. When I was young, I used a ripsaw, the advantage of which was that you could follow the grain direction and character of the wood. It was a satisfying feeling when you managed to fit all the parts in the right place. Today, with machines, there's no trouble sawing to the line, but during traditional times there was only 1/8" (3mm) more thickness than the finished material, so you had to be dexterous and careful to saw the face of the board almost perfectly perpendicular.

After ripping all the framing material, start preparing the hipboard. Commonly, the figure in the hipboard runs vertically, so 99 per cent of the time you won't have a one-piece hipboard. You might have to glue three or four pieces of wood edge to edge. Commonly you can buy 3/8" (9.5mm) thick hipboard material. However, if this thickness is not available, then we saw a 7/8" (22mm) thick board into two pieces. Traditionally, the wood is sawn by hand, though contemporary craftsmen use a bandsaw (photos 7-10).

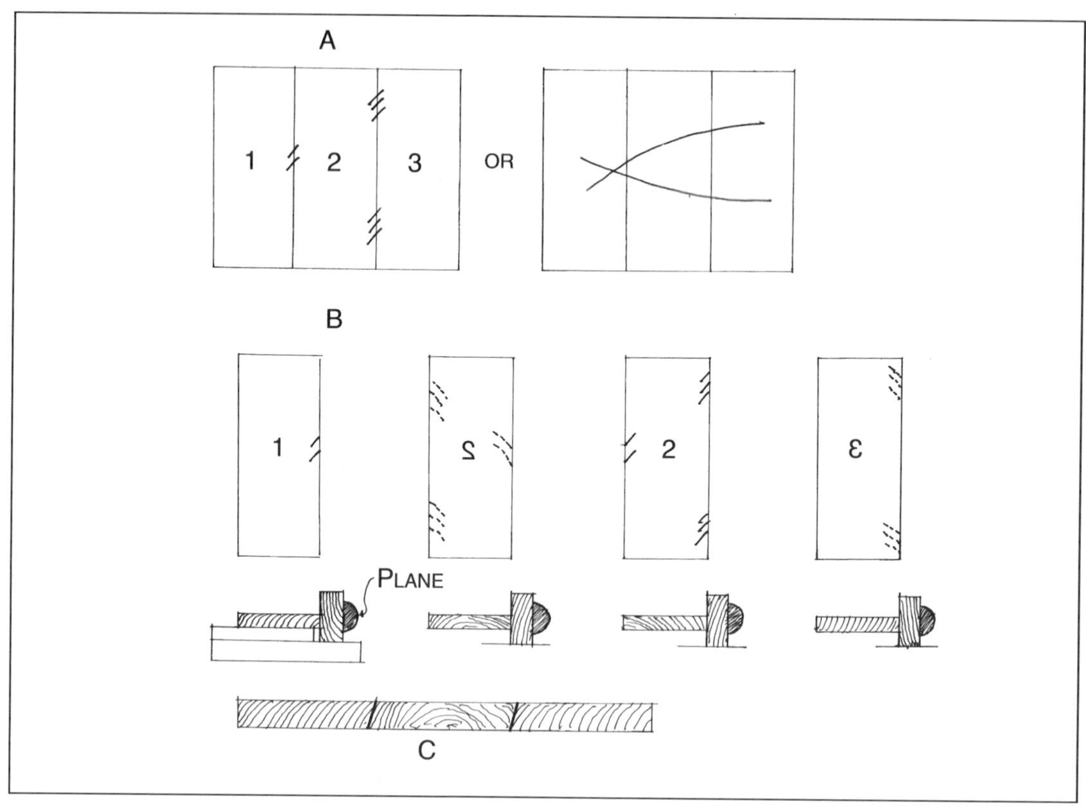

D5 Mark how the boards go together, but flip the center board over before planing the edges. This way, the glued hipboard will be flat even if the edge is planed slightly off square.

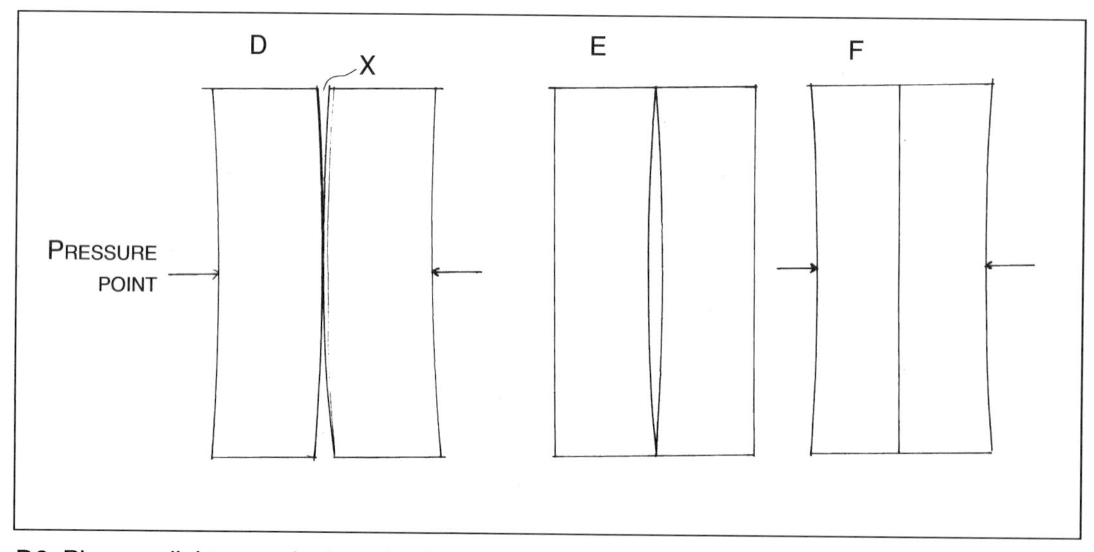

D6 Plane a slight curve in the edge by pressing harder in the middle of the stroke.

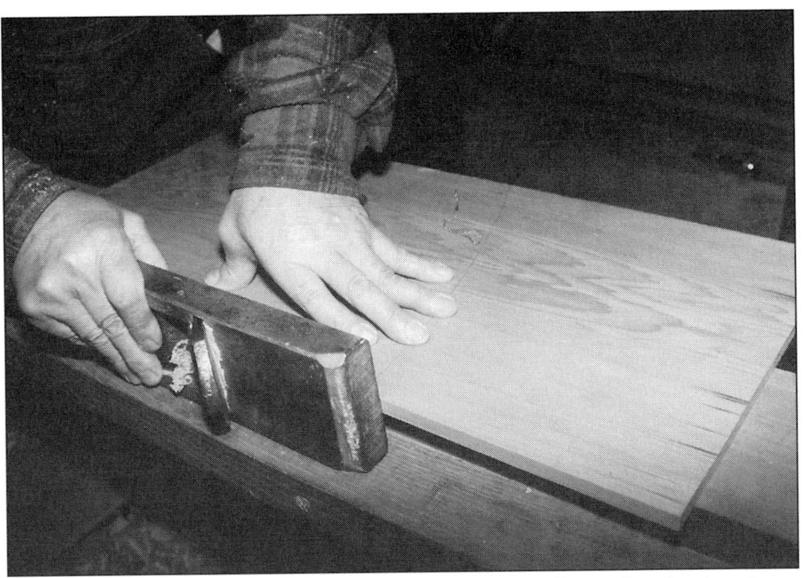

11 Use truing plane, which has a specially configured sole, to plane edges of hipboard before gluing the boards together. To create a slight curve on the edge, press lightly when starting to plane, press harder when reaching the middle, and release pressure gradually at the end.

12 Before planing the edges place the three boards together and mark one X sign in the center, thus you will remember the original position of each board.

GLUING BOARDS EDGE-TO-EDGE

A *tategu-shi* (sliding-door maker) uses a unique method to edge-glue a thin board. The method is called *hagi* (gluing boards edge to edge), and it is commonly used for softwood up to 1/2" (12mm) thick. As an example, suppose you were gluing a three-piece hipboard. First, lay the pieces down and decide how to place them together. Next, mark their position. I mark them as shown in drawing 5, label A (photo 12). Then plane the edges as shown in drawing 5, label B, flipping the center board over (photos 13-14). This way, if the edge is planed slightly off square, the board will still be glued flat (see D5, label C).

Planing these edges is a bit difficult because if you simply plane the edges straight, both ends of the faces may open when being glued (D6, label D). So you must calculate the pressure and the opening (D6, label E). To create a slightly curved edge while using a truing plane, press lightly when starting to plane, press harder when reaching the middle, then gradually decrease the pressure as you reach the end.

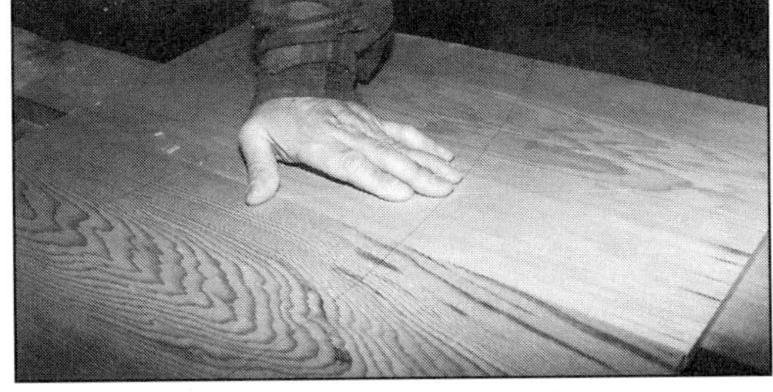

13-14 Checking alignment of edges.

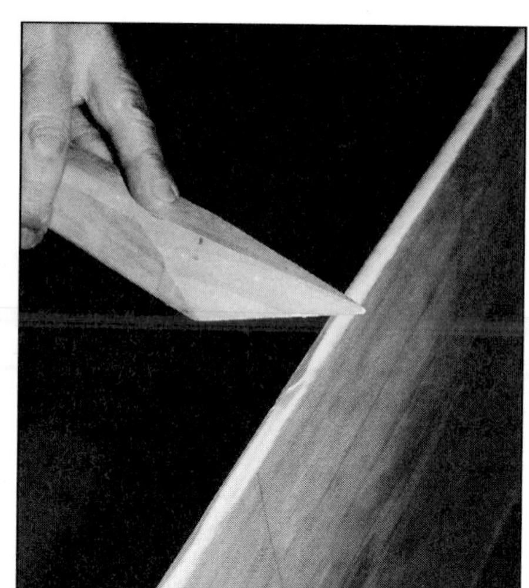

15-17 Gluing edges of boards.

18 Tying boards together with a rope, edges should tightly touch each other.

19 Using pieces of wood to give pressure to the boards.

20 The board is glued, tightly pressed, and is now drying.

To glue the pieces tightly, traditional craftsmen use a rope with a pair of 2x3 battens in the center of the boards. Apply pressure around the middle of the boards by inserting two short pieces of wood in between the 2x3s and the rope. These will put the rope into tension, like a guitar string stretches over a bridge. This way of proceeding will put uniform pressure all along the glued edge (D6, label F). To learn how to make and apply traditional rice glue, see the chapter entitled "Japanese Rice Glue," page 86.

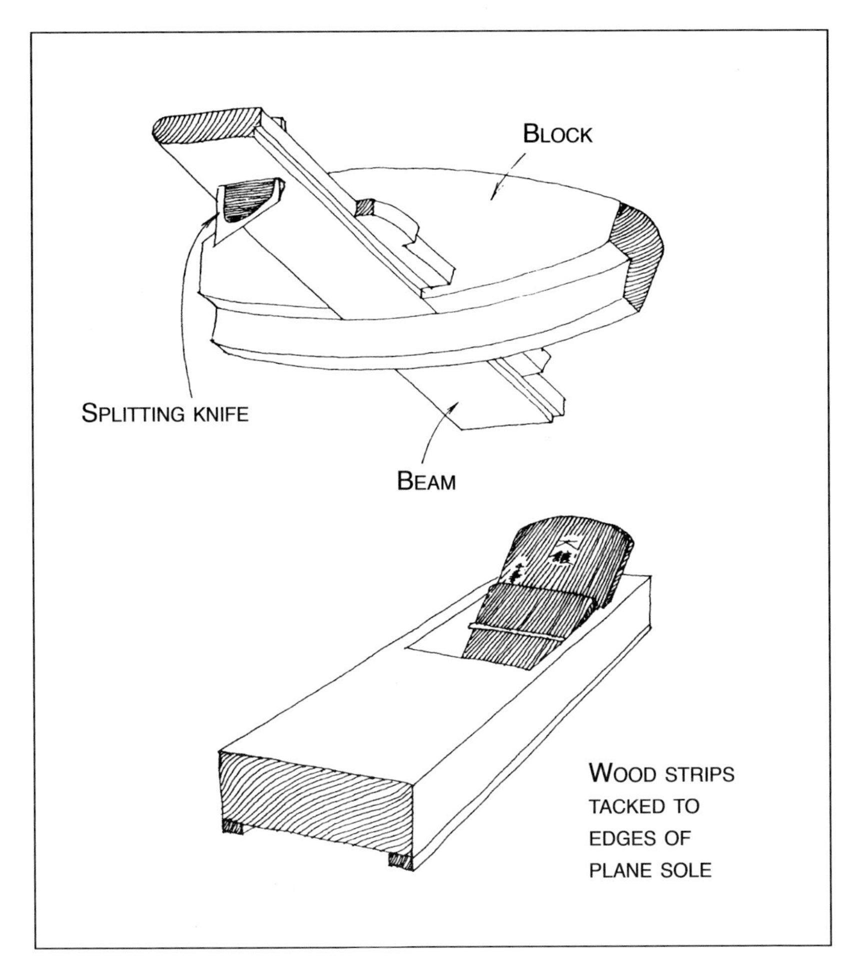

D7 The splitting gauge and plane with depth-regulating strips make *kumiko* quickly and accurately, and minimize loss of wood.

SPLITTING THE *KUMIKO*

While the glue is drying, I prepare the *kumiko*. Today you don't have to follow traditional practices, but they are still interesting to know. Traditionally, we would first plane one side of the 7/8" (22mm) board completely flat, checking across the grain with a straightedge to make sure there was no cupping. Then we would use the marking gauge from the planed side and mark 5/8" (16mm) to 3/4" (19mm) on both edges and ends. We would plane the wood down to the marks while continuing to check with the straightedge. The final thickness was between 5/8"(16mm) and 3/4" (19mm)—the decision was up to the craftsman. The next step was to use the truing plane with support (D7). We would plane the edge square to the face, then use a splitting gauge set to the thickness of the *kumiko*. By deeply scoring the wood from both sides, we could snap the *kumiko* off the board.

The planing and splitting process was repeated until all *kumiko* were cut, making sure to cut plenty of extras. Then we soaked a cloth in water and saturated the compressed side to relieve the pressure made by the splitting gauge. If you do not saturate the *kumiko*, they will eventually swell, causing the entire assembly to warp.

21 Plane the edge of the board square before splitting another kumiko.

22 The splitting gauge has a curved block, which allows it to cut deeper with each pass.

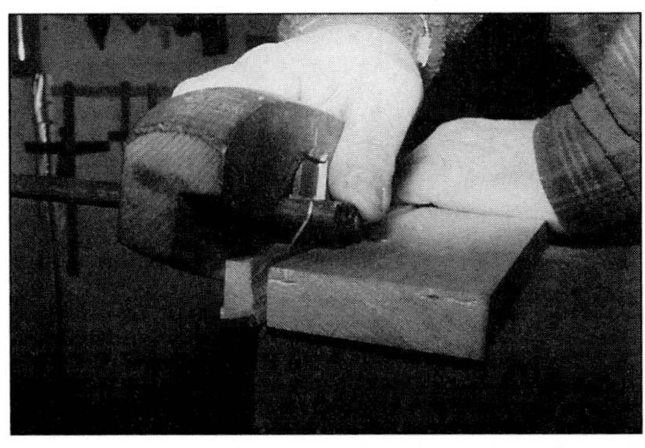

23 The splitting gauge breaks a kumiko off the board.

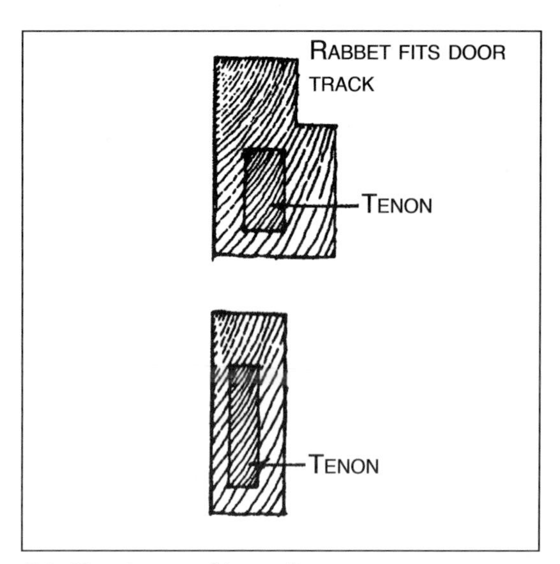

D8 Two types of top rail.

24 Tacking the depth-regulating *kumiko* strips onto the sole of the plane.

25 The strips act as a thickness gauge when planing the split *kumiko*.

PREPARING THE RAILS AND STILES

While the *kumiko* are drying, I prepare the stiles. The front faces of the stiles will receive the paper, so they must be planed flat and free of twist. Next, I plane the inside edge square to the front face. But instead of being straight along its length, the wood should bow slightly to hold the stile tightly against the *kumiko* shoulders. What keeps the structure in place are the large tenons on the rails, which are made to fit tightly into the stile mortises. Since the *kumiko* tenons are small and delicate, they are primarily used for positioning, not for strength.

Once the front face and inside edge have been planed, gauge the width of the face with a marking gauge and plane the outside edge. Then gauge the thickness of the edge and plane the back face. This is the face that will show in the room. All the frame parts of the *shoji* are planed in this order. All the rails are about 1/16" (1.6mm) thinner than the stiles. This way, without disturbing the shoulder, you can chamfer the inside corner of the stile's back face at the final planing.

Here I will illustrate two customary types of top rails (D8). Which one to choose depends on the craftsman and customer. I will elaborate on this further on.

At this time the *kumiko* would be dry. With a special plane, I would simultaneously plane two or three *kumiko* to the desired thickness (D7, page 21). The plane had a *kumiko* on both edges of the sole to regulate thickness. My master and other old timers didn't have *kumiko* planes when they were young. He could make even thicknesses by touch—he would just feel the *kumiko* with his fingers. It was amazing to see. However, as he and the other old timers became even older, they did use a special *kumiko* plane that had two springs and a roller. It pressed down the *kumiko*, and the blade would cut. Today almost no one uses this skill any longer.

26 Cut the hipboard to a rough height.

27 Mark the thickness of the hipboard with a marking gauge.

PLANING THE HIPBOARD

At this point, all *kumiko* preparations will be complete, so go back to the glued-up hipboard. Commonly, the tool that makes the groove in the frame, called *kikai-shakuri kanna* or plow plane, comes in different sizes. The most common groove sizes are 3/16″ (5mm) and 1/4″ (6mm), but 1/8″ (3mm) and 5/16″ (8mm) sizes are also available. Make the thickness of the hipboard match the width of the groove. First you roughly plane one side of the hipboard and cut the board slightly larger than the final piece. Mark the thickness of the board by placing the marking gauge against the planed side, making sure to make it a little thicker than the final thickness. Then plane the other side with the medium finishing plane to a clean finish at final thickness.

28-29 Medium finish planing. When planing hipboards, use a wider plane. If you use a narrow plane, the convex blade will make many seams in the wood.

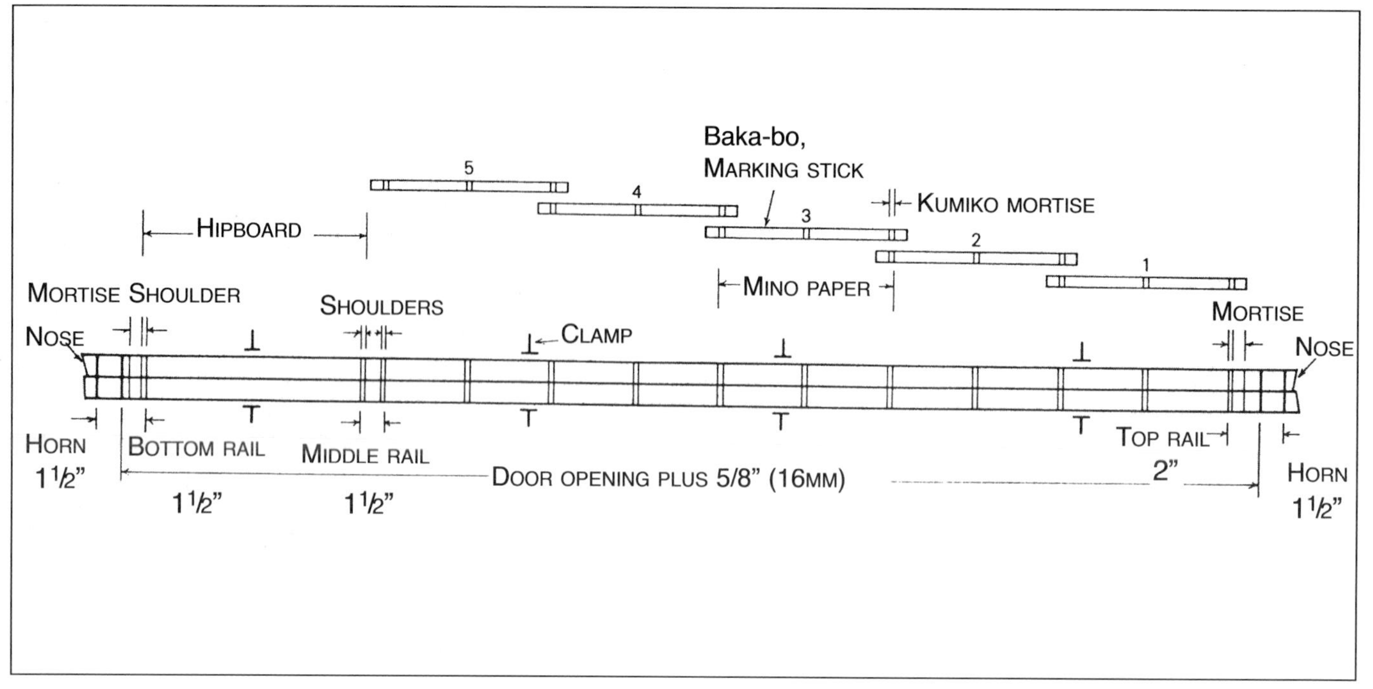

D9 For a pair of *shoji*, clamp four stiles together and lay them all out at one time. Lay out kumiko mortises from the top rail, using a marking stick. Space the *kumiko* according to the width of the *shoji* paper, and adjust the height of the hipboard to make the *shoji* fit its opening.

LAYING OUT THE JOINTS

The wall openings and tracks built by the house carpenter determine the outer dimension of the *shoji*. The width of the *shoji* paper determines the spacing of the horizontal *kumiko*. Marking out this spacing from the top rail also determines where the middle rail will go. This determines the height of the hipboard. All other measurements depend on the discretion of the craftsman. For speed and accuracy, lay out the same or similar pieces (both stiles for instance, or four when one opening requires two *shoji*), at the same time. Use clamps to align the pieces. The *tategu-shi* uses clamps mostly for layout and almost never for assembly.

First I mark the stiles (D9). Don't forget that each has a right and a left. Since it is customary to orient the wood in the stiles in the way in which it grew in the tree, make sure that the largest growth rings are at the bottom of the stiles. Clamp the stiles with their inside edges up and mark. Starting from the bottom, mark first with a pencil about 3/8" (10mm) to 1/2" (13mm) up. This is the nose, which you will cut off in order to square the end of the wood. From this line make a pencil mark 1" to 1 1/2" (25mm to 38 mm) up, for the horn. Next, use a marking knife to establish the width of the bottom rail. From this knife mark, mark down 1/4" for the cosmetic shoulder line. This also indicates the depth of the groove for the

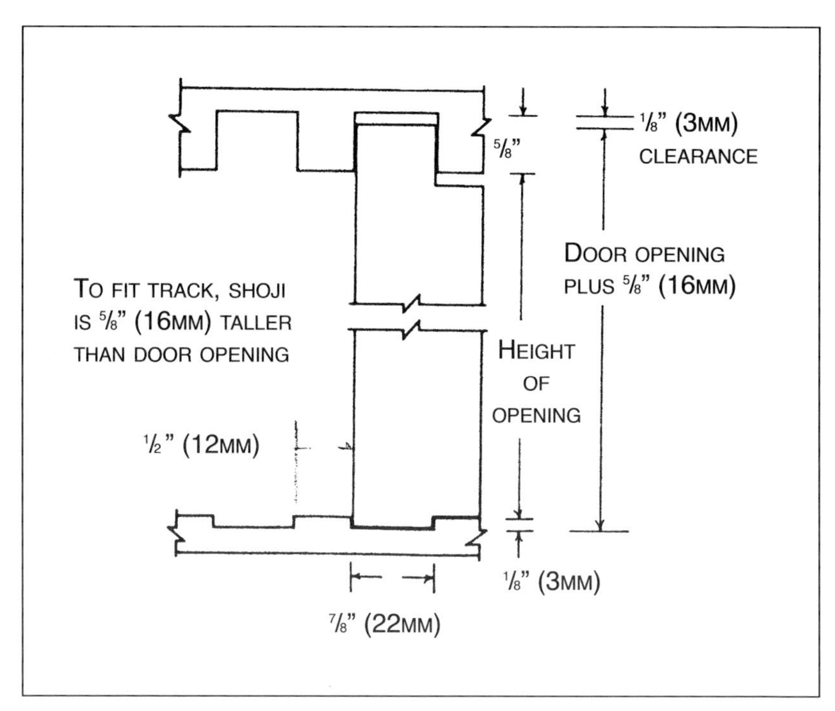

D10 Determining height of *shoji*.

30 Mark mortise position on stiles. Clamp the stiles together so you can transfer the layout marks across them all at once.

hipboard. Mark down the mortise height (photo 30). Carry these marks all across the width of the stock with a square.

Next, mark the height of the *shoji*, measuring from the bottom line of the bottom rail to the top of the opening plus 5/8″ (16mm). This extra length fits the top and bottom tracks in which the *shoji* will slide (D10). From this pencil mark, mark another pencil line 1″ to 1 1/2″ (25mm to 38mm) away, which will be the upper horn. Then mark down from the height of the *shoji* mark to the bottom edge of the top rail and then measure up 3/16″ to 1/4″ (5mm to 6mm) for the cosmetic shoulder line. Finally mark the height of the mortise.

The horn, which is also found in Western joinery, has several purposes: When you assemble *shoji* or any other door, much stress is applied on the top and bottom mortise-and-tenon joints. If there were no horn, the joint might split during assembly. Horns also prevent accidental damage. If you keep the sharp corner of the end grain, you might hit the floor and chip the wood.

After assembling the *shoji*, check for twist and squareness. If the house column is not square to the bottom track, the horn allows you to compensate. Without the horn, you would have to plane the outer edge of the stile, which you cannot do because the tenons are in very deep.

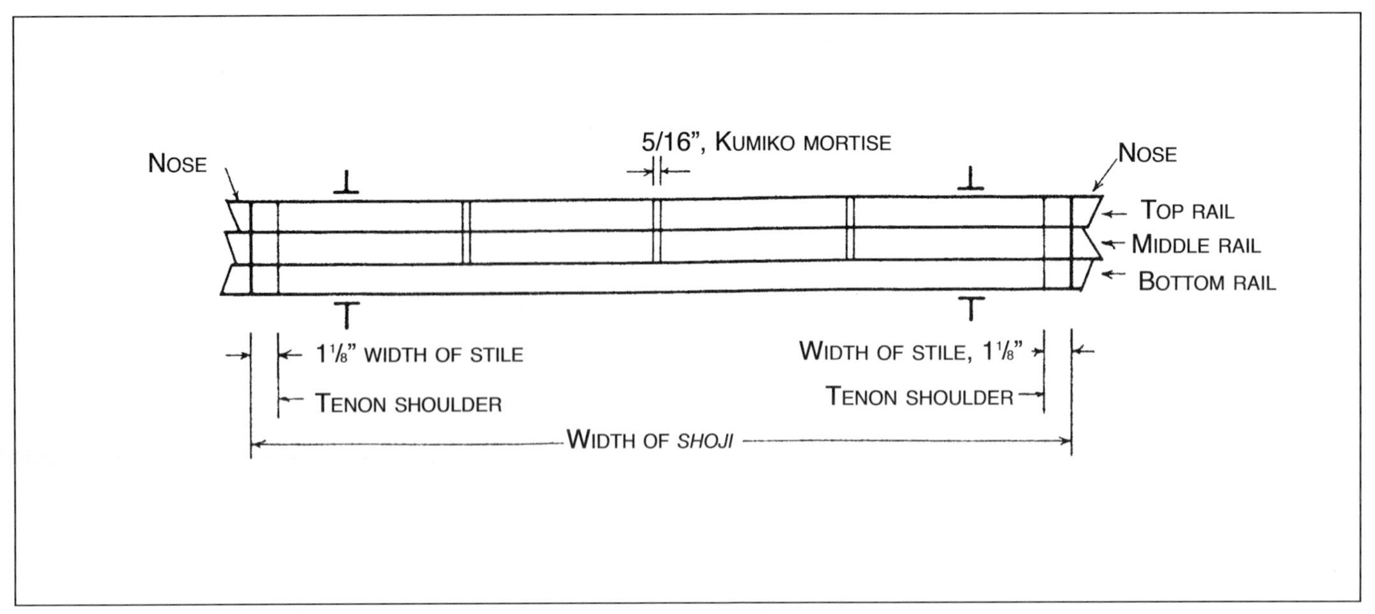

D11 For a pair of *shoji*, lay out six rails at once. Width of shoji (length of rail) equals width of opening plus width of one stile, all divided by two. Mark *kumiko* mortises on top and middle rail only. Space between *kumiko* equals distance between stiles, minus combined width of *kumiko*, all divided by number of spaces.

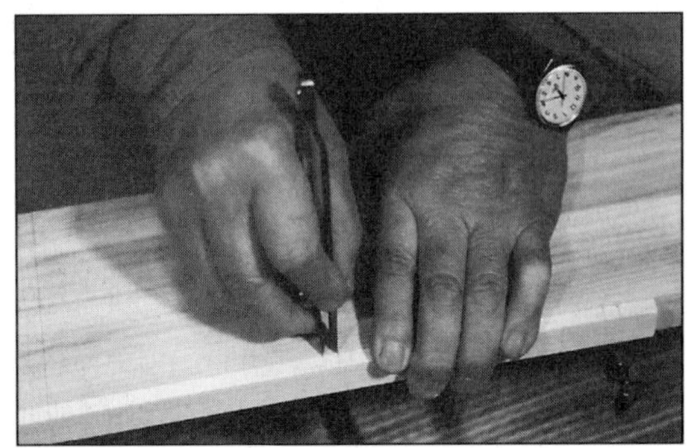

31 On stiles, trace from the marking stick width of the *kumiko* mortises, and the distance between mortises.

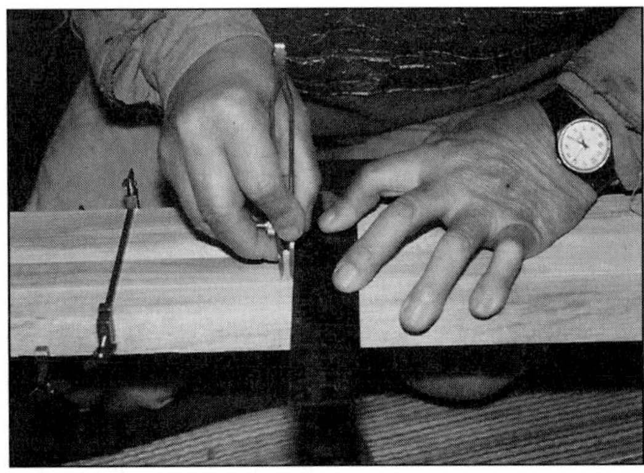

32 After tracing from marking stick, use square and mark the whole width of the stiles with marking knife.

MARKING METHOD FOR *KUMIKO*

The spacing between horizontal *kumiko* depends on the size of the *shoji* paper. There is a common method for marking the *kumiko* mortises accurately using one piece of *kumiko* material to make a marking stick called *baka-bo,* which roughly translates into "foolish stick." I call it the "smart stick," because if you were to measure each space with a ruler, you would create multiple errors. The marking stick makes the job faster and more accurate.

The marking stick represents one length of *shoji* paper. It carries the width of the paper and the position of three *kumiko* in relation to that width. D9, 11 shows the layout of *kumiko* mortises for *mino*-size paper, which gives nine horizontal *kumiko*. The stick has two *kumiko* mortises marked just inside the paper width, plus one centered between them. All *kumiko* marks are drawn with a double marking knife, which is set for the width of the *kumiko* face, or just slightly less (photo 31). Beginning at the top rail mark, overlap it with the width of one *kumiko* mark. I match the inside of the first *kumiko* mark to the bottom line of the top rail and knife off the other marks. Next, I reposition the stick to overlap the last *kumiko* mark and mark the next two marks, and continue in this manner five times. Then, inside of the fifth or last *kumiko* mark will be the top middle rail. Now I have nine *kumiko* marks. Finally, I mark the width of the middle rail and within it a 1/4" (6mm) mark indicating the top and bottom cosmetic shoulder and the hip-

33 Clamp six rails together and mark the width of the stiles at both extremities with marking knife.

34 Mark the *kumiko* mortises on only 4 rails, that is, the top and middle rails. These rails will have three vertical *kumiko*.

board groove. Between these marks will be the height of the mortise's middle rail. I square all these knife marks across all four stiles (photo 32), saw off the noses (the waste beyond the horns), unclamp the stiles, and chamfer all the ends to prevent damage.

Next will be the rails. When two *shoji* fill a door opening, they overlap each other by the width of a stile. I mentioned earlier that the width of each *shoji* is calculated by adding the width of a stile to the width of the opening and dividing by two. To make the rails, I clamp them together, inside edges up, and mark the nose with a clean line about 1/4" (6mm) from the end. If necessary, you can mark more deeply with a square and knife. Then I mark the length of the *shoji* width with a knife and square. Next I mark the width of the stiles from both ends with a knife (photo 33). These lines locate the tenon shoulder (photo 36).

Mortises for the vertical *kumiko* are marked next, equally spaced between the two stiles. Use the same double marking knife and a new marking stick, which has only two *kumiko* marks (photo 34). The rail has only four spaces, so if you think you can divide it evenly, then you don't have to make the stick. After finishing the markings, cut off the noses squarely. Squareness is important for marking tenons (photo 35). Now unclamp the rails and extend the shoulder lines, marked on the inside edge, all around (photo 36).

35 By placing the stock on both horses, saw squarely ends off noses.

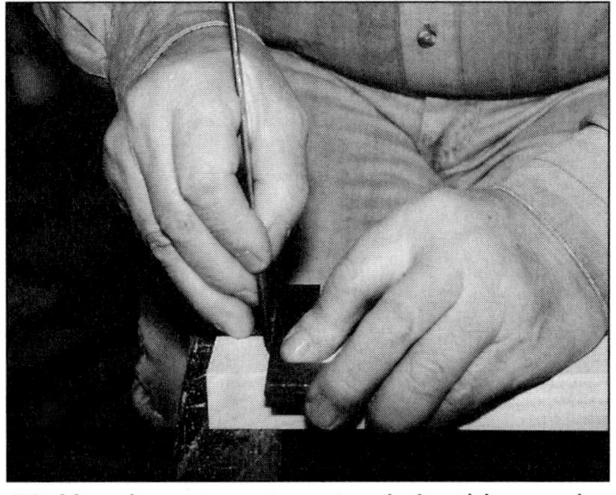

36 Use the square to extend shoulder marks of tenons all around.

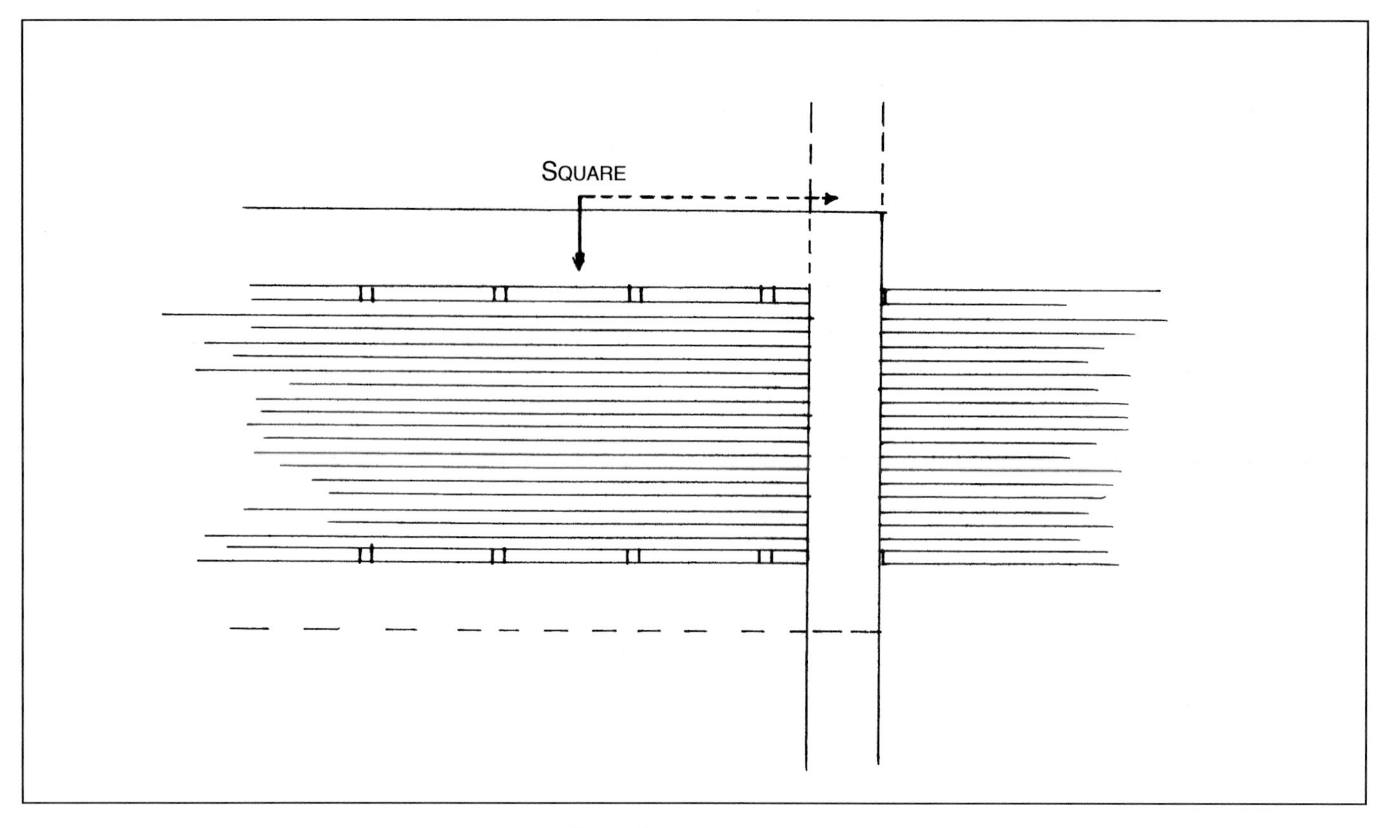

D12 Place unmarked *kumiko* in between two marked *kumiko*. Use a square to make sure they are lined up. Then you will be able to transfer the marks to all the unmarked *kumiko*.

MARKING *KUMIKO* AND RAILS

37 Choose seven *kumiko* to clamp onto stile then with a knife trace the seven *kumiko* mortise marks from stile.

Next mark the vertical *kumiko*. To lay out the tenons and mortises for the half-lap joints on the vertical kumiko, transfer the layout lines from one of the stiles to two of the kumiko. Clamp them and use the same double knife. From these two marks, transfer the marks to the rest of the *kumiko* (D12). Earlier I had indicated with a mark the bottom of the stile; also transfer this mark onto the *kumiko* with a pencil. This is a reminder for the bottom of the *kumiko*. Later, transfer these marks onto the tenon.

When you clamp the two marked *kumiko* on both extremes of the unmarked *kumiko* stack, make sure the two marked *kumiko* are lined up squarely to each other. It's a good idea to mark two extra *kumiko* and not use the marked *kumiko* in the finished *shoji*. Transfer the marks onto the rest of the *kumiko*. The vertical *kumiko* get notched alternately front and back, so I square-mark every other notch around the underside of the stack. Mark all notch-marks on both edges, mark tenon shoulders with a knife all around, and mark tenon length with a pencil. Also mark the depth of notches by using a marking gauge set a hair deeper than halfway (photo

38 After transferring all needed marks on *kumiko* from stile, unclamp stiles from *kumiko* by taking off one *hatagane* clamp at one end while keeping the *kumiko* unmoved and in a group at the other end.

39 Reclamp *kumiko* separated from stile. This method is always used to keep matching parts aligned, for both marking and cutting.

40). This is the common way to lay out *kumiko*.

If the *kumiko* number fewer than ten, you can clamp the whole stack to the stile and transfer the layout lines (photo 37). When you are transferring the notches from the stile, you can't mark all of them across the stack, so you'll have to skip every other one, except the first and last *kumiko*, which should both be marked. Having marks on these two *kumiko* helps you when marking the other side of the stack. Again, it's a good idea not to use the two outside *kumiko* in the finished *shoji*. Then unclamp all the *kumiko* except the second from the last. Prepare the opening of some free clamps to clamp only the *kumiko*, then lift the unclamped *kumiko* ends very gently untill they're clear of the stile—just enough to be clamped alone. Clamp the *kumiko* and unclamp the last one to separate the *kumiko* gently from the stile (photo 38). Quickly clamp the *kumiko* and secure them by keeping the *kumiko* unmoved in a group (photo 39). Now you can mark the unmarked alternate notches on the reverse side, and complete the *kumiko* as previously described.

40 On the vertical *kumiko*, mark depth of lap joint with marking gauge a hair deeper than half-way through. It is better to avoid marks on the outside of the notch, because once assembled one could notice them.

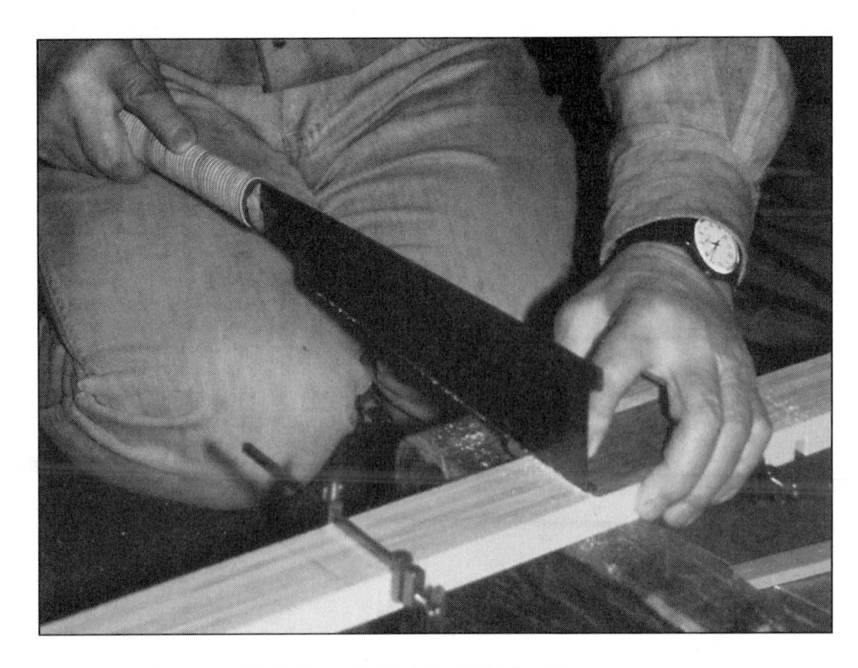

41 Sawing the lap joints of the clamped *kumiko*, the first quarter-inch with the help of a wooden jig, then I continue freehand.

42 To break out the waste, I pull the corner of a flat chisel along the kerf.

43 Then I clear the waste with a mortise chisel run in the notch, bevel-side down.

CUTTING THE JOINTS: VERTICAL *KUMIKO*

While the vertical *kumiko* are still clamped together, saw the notches and tenons. First cut both shoulders of one notch, making sure to cut inside the knife mark and straight down to the depth line. Use a straight piece of *kumiko* material or a wooden square as a cutting guide (photo 41). Break out softwood waste with a chisel (you can't do this with hardwood *kumiko* material), and clean up by running a mortise chisel, bevel-down, along the bottom of the notch (photos 42, 43). Make sure depth is uniform. Then insert a planed-to-fit scrap of *kumiko* into the notch. The notches are a hair narrower than the *kumiko*, but before assembly you will plane the *kumiko* to fit. The planed scrap *kumiko* is placed in the notch for security in case a clamp shifts while cutting the other notches.

Next cut the *kumiko* tenons. Cut the length of the tenon squarely. This time the bottom indicator pencil mark is gone, so remember to mark the bottom side of the tenon. Mark the tenon on the end grain of the stack with a 1/4" (6mm) marking gauge (photo 44). Mark dead center and on the edges of the two outside *kumiko*. After finishing the *kumiko*, don't break the gauge's set, for you will use it on the rails and stiles. Saw the shoulders using the same straight stick or a wooden

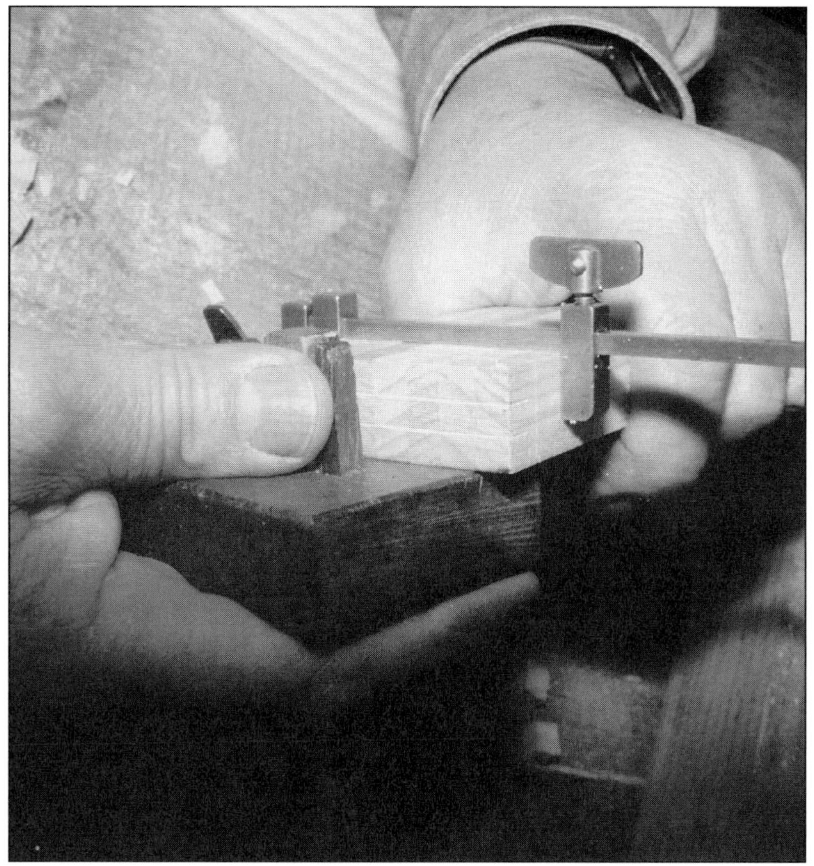

44 The tenons are marked on faces and end grain with a marking gauge.

45-46 You can saw the shoulder of the *kumiko* tenons, then break off the waste with a chisel pushed in from the end grain or, as I am doing, use a *kiwa kanna* by shaving material from both sides of the *kumiko* to the tenon thickness line.

square to guide the saw. These are small tenons, so instead of ripping to meet the shoulders, I use a chisel to break off the waste from the end grain. My index finger on the underside of the chisel acts as a stop to keep the chisel from damaging the *kumiko* shoulders. In all but straight-grained stock, I break a little bit wider and pare the tenons to the line. You can also plane the tenons to shape. The Japanese have two or three different kinds of rabbet planes for this. I use a special plane called *ai-jakuri-kiwa-kanna,* or rabbet-cutting plane (photos 45-46). You take cross-grain shavings to cut down to the line. When there are numerous *kumiko* to cut at the same time, planing in this way is better. However, it might be a bit clumsy when you plane just a small number. This plane is also good to use when the wood grain isn't straight.

After finishing the tenons, I put a pencil mark on the cheek to indicate which tenon is the bottom one. This way the marks won't disappear during finish-planing. After all the notches and tenons are cut, and before removing the clamps, I chamfer the upper and lower edges of the tenons (photo 47). Then I remove the clamps and fan out the stack to chamfer one corner then the other (photo 48). The vertical *kumiko* are now ready.

47 When the tenons have been formed, chamfer them at a 45 degree angle with a few strokes of a plane.

48 Then unclamp the stack, fan out the *kumiko*, and chamfer the other two corners.

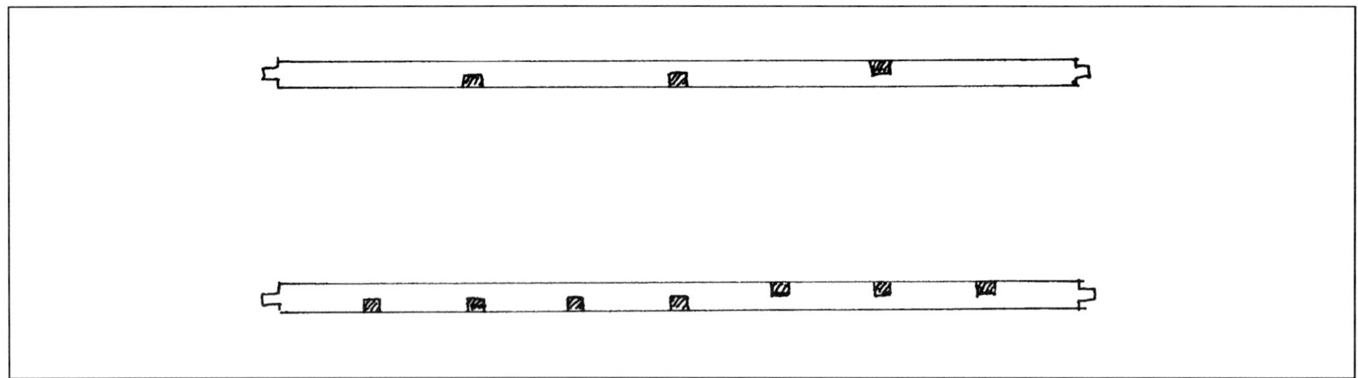

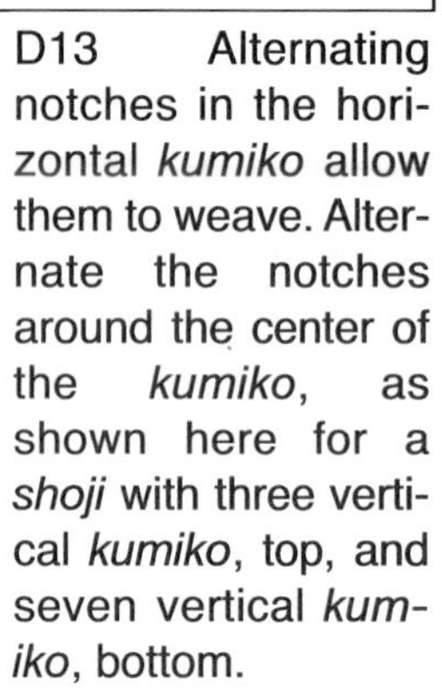

D13 Alternating notches in the horizontal *kumiko* allow them to weave. Alternate the notches around the center of the *kumiko*, as shown here for a *shoji* with three vertical *kumiko*, top, and seven vertical *kumiko*, bottom.

49 Once the vertical *kumiko* have been finished, work on the horizontal *kumiko*. Detach marked horizontal *kumiko* from rail and sandwich in between them all other 18 *kumiko*.

50 Before marking, verify the material's flatness by hammering the *kumiko* flat then feeling with your touch if more needs to be hammered or just pushed with your fingers.

51 With a square match top and bottom marks and trace all other sandwiched *kumiko*. Extend the marks further on both sides and back.

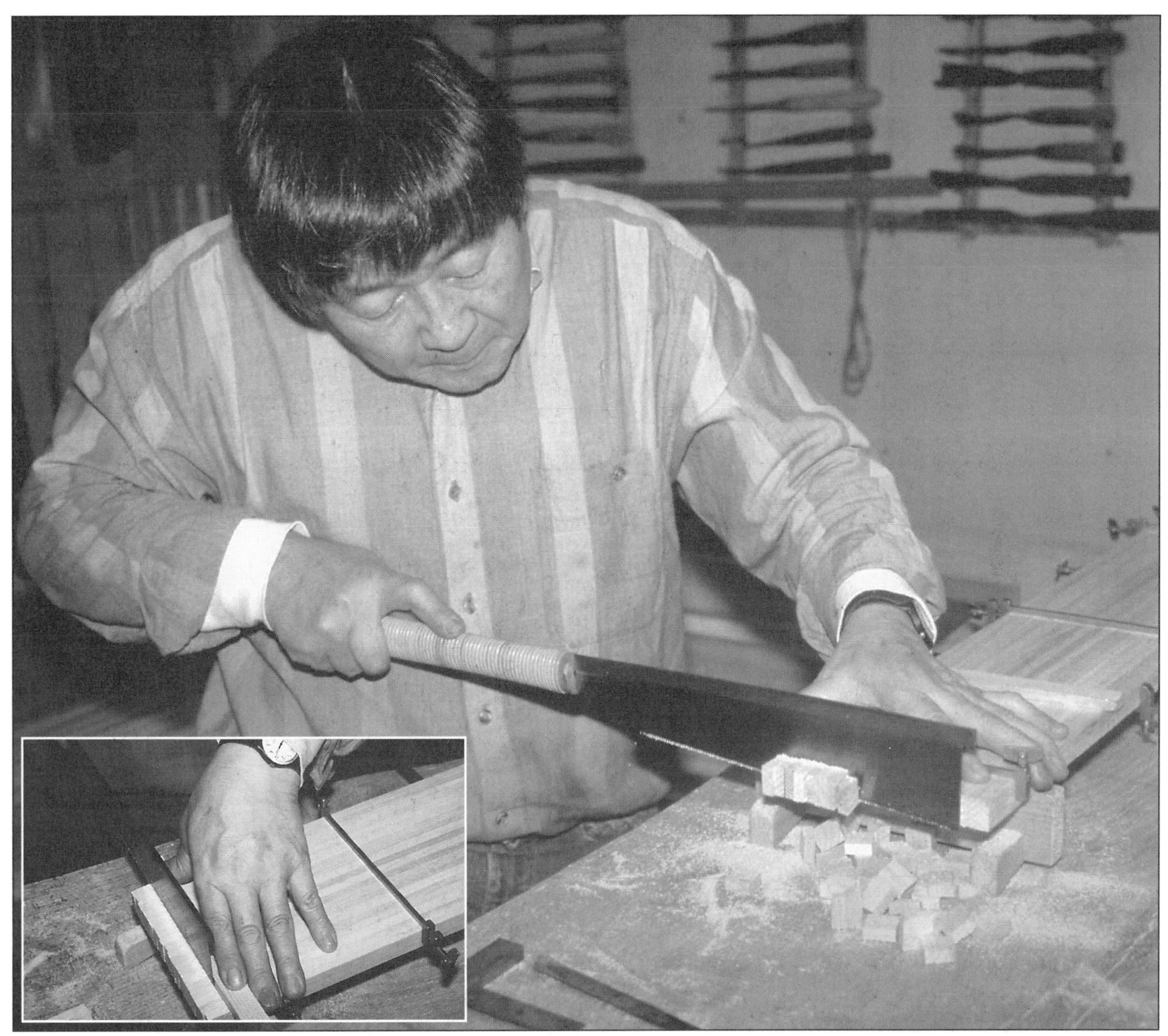

52-53 Cut nose square for end of horizontal *kumiko* tenons.

CUTTING THE JOINTS: HORIZONTAL *KUMIKO*

Many people think that the *kumiko* overlap every other one, as in a woven basket or half-lap joints. But because the *kumiko* are not flexible, the Japanese had to create a way to lock the pieces together. When there are three vertical *kumiko,* for instance, the notches in the horizontal *kumiko* are two adjacent on one face, one on the other. And if there are seven vertical pieces, then the horizontal *kumiko* have four notches in front and three on the back to prevent the screen from warping (D13). They are marked out and cut exactly like the vertical *kumiko* (photos 49-54). Regardless of the number of horizontal pieces, vertical notches are always cut alternately.

54 Detach *hatagane* and fan horizontal *kumiko* to chamfer tenons.

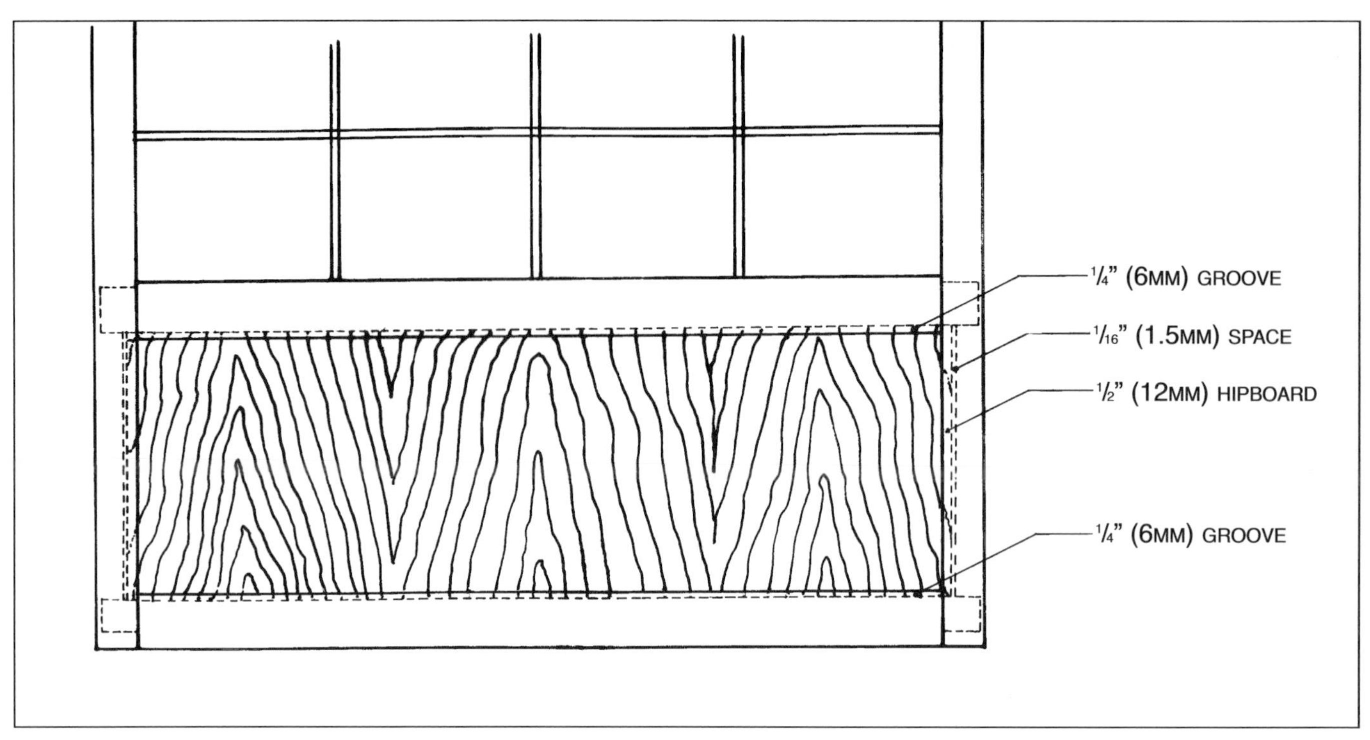

D14 Position of stiles, rails, and hipboard.

MARKING THE FRAME

55 Mark tenon on ends of rail: bottom and middle rail tenon is 3/8"(9.5mm) thick, top rail tenon is 1/4" (6mm) or 3/8" (9.5mm) thick.

Next comes marking the mortise width on the stiles. Gauge the mortise width, and make sure the fence is always on the front face of the stack. First decide on the position of the large mortise and groove: Should they be in the center of the stile, or a little to the left or right? Since the hipboard groove is involved at the joint, the mortise will be 3/8" (9.5mm) wide. The groove is commonly 1/4" (6mm) wide. Centering both the groove and the mortise will weaken the joint, so you must try to separate them. But moving either one too much off center will decrease strength, and, in the case of the hipboard joint, compromise aesthetics as well. Thus the craftsman's judgment on placement is crucial. I mark the mortise a little bit to the front and the groove a little bit to the back. This way the mortise and groove will not meet head to head. This takes care of the bottom and middle joint on the stiles.

Next is the top rail. As I mentioned earlier, there are two types of rails. If you choose the full-thickness type, the mortise will be 3/8" (9.5mm), but if you choose the 3/4" (19mm) rail, the mortise will be 1/4" (6mm). Some craftsmen use double mortises with a full-thickness top rail. Both are 1/4" (6mm) wide, but have different heights—because the back of the rail is rabbetted, the front top is higher. The craftsman's judgment on placement is critical here, too. Once you have decided on the mortises, mark them on the stiles. With the same mortise marking gauge, mark joints on both the edges and ends of the rails (photo 55).

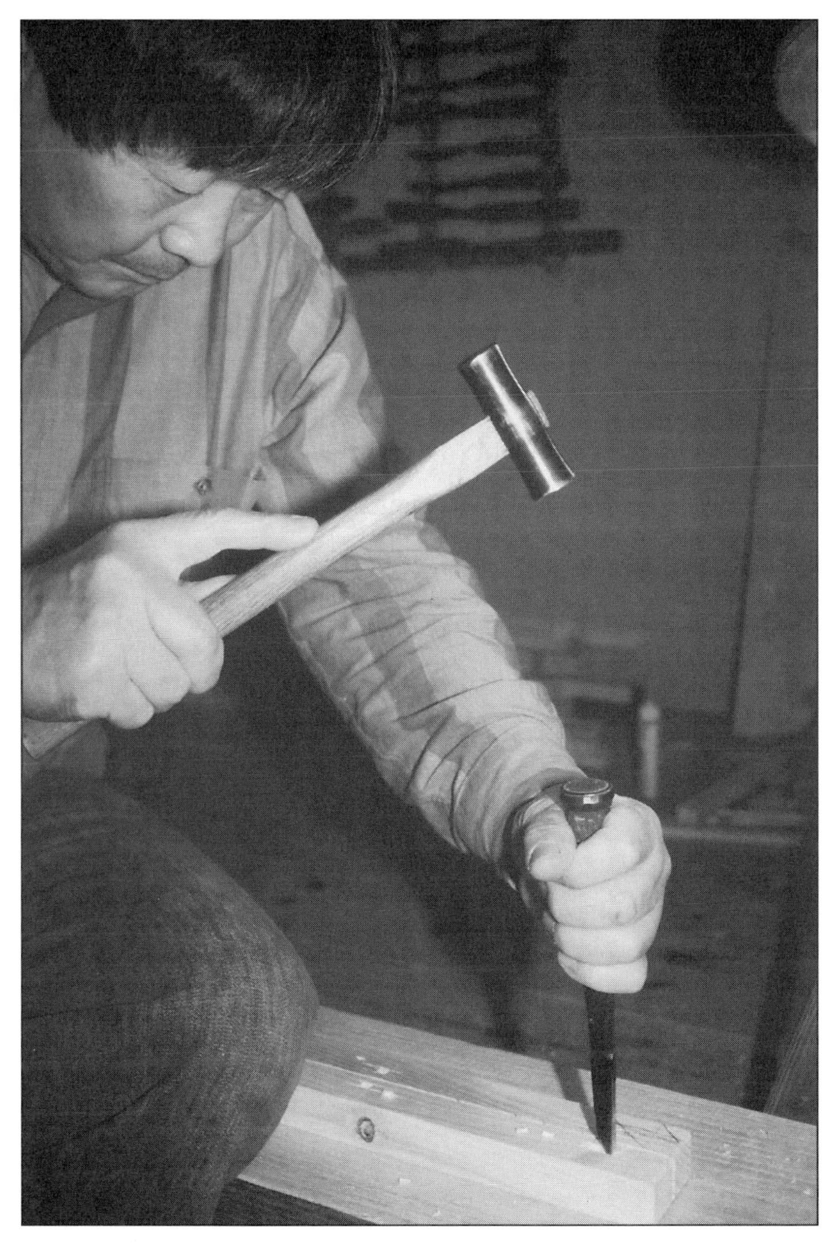

56 The *tategu-shi* sits on the wood to steady it while mortising it. He places his chisel frequently in a bamboo pot of cotton wadding soaked with vegetable oil, to reduce friction. The chisel has three concave sides and a hollow-ground face. He chops from the middle out, always with the face toward the middle of the mortise, except for the final cuts at either end of the mortise. These are angled slightly from the perpendicular to taper the mortise for a tight fit when the tenon is driven in.

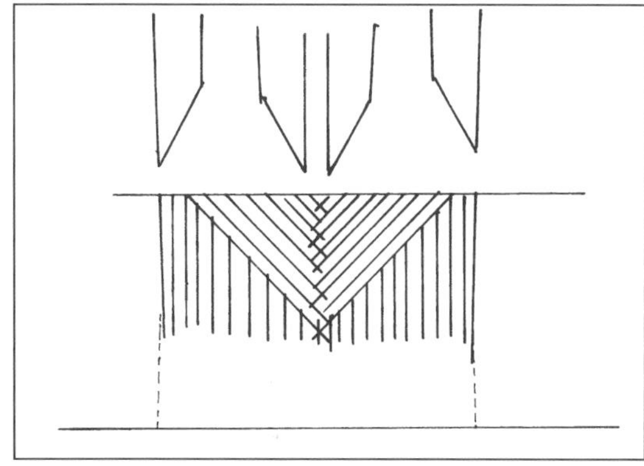

D15 Chop the mortise from the middle out, always with the face of the chisel toward the center.

57 To reduce friction while mortising, dip the chisel into cotton wadding soaked with vegetable oil.

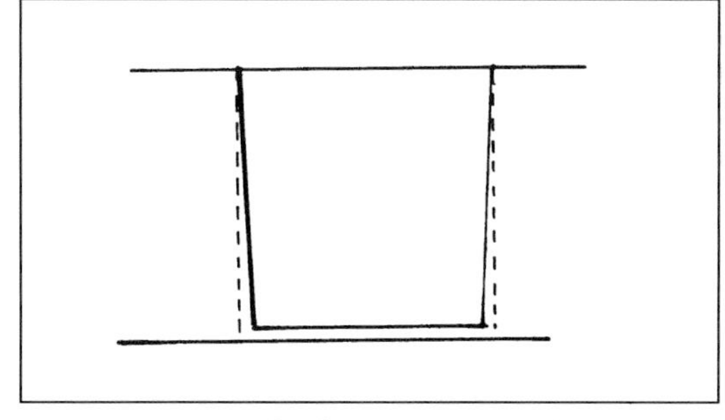

D16 Angle the final cuts to taper the mortise slightly, for a tight fit.

Kumiko mortises are marked with the same marking gauge that you set to mark the *kumiko* tenons. The *kumiko* notches and tenons were cut while the *kumiko* were clamped up for layout. The joints on the stiles and rails are now cut individually, mortises first, then tenons. Cutting the tenons last lessens the danger of damaging them. In Japan, the quality of a craftsman's skill is judged by his speed and accuracy. It is considered most important to make each saw or chisel cut the final cut—you go directly to the layout line. The less contact with the wood, the less chance of error and the more crisp the finished work.

The mortise width mark is exactly the width of the mortise chisel. All *shoji* mortises are blind. See the process of mortising in the chapter "The Japanese Mortise-and-Tenon," in photos 56 on this page, and 58-63 on page 36.

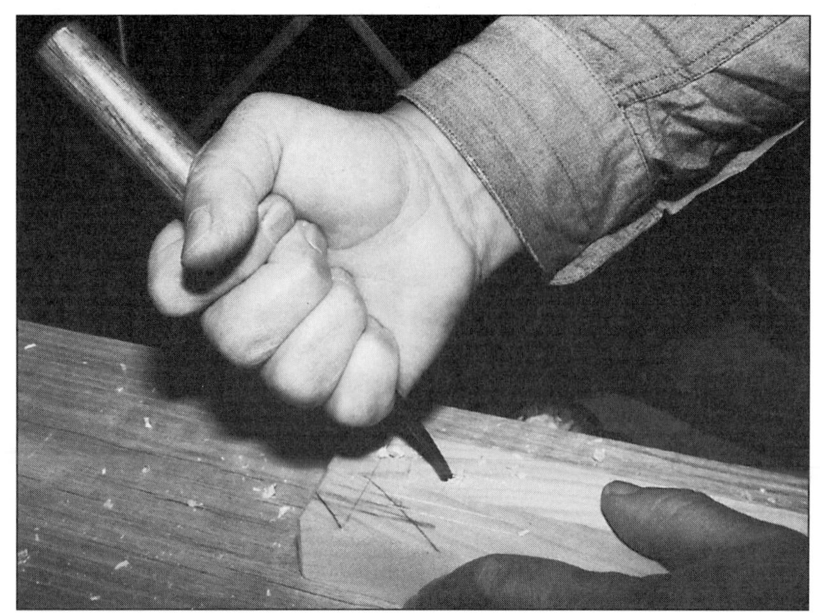

58 Scraping the bottom of the mortises with the *sokozarai nomi* (bottom-cleaning chisel) at the top of the stile.

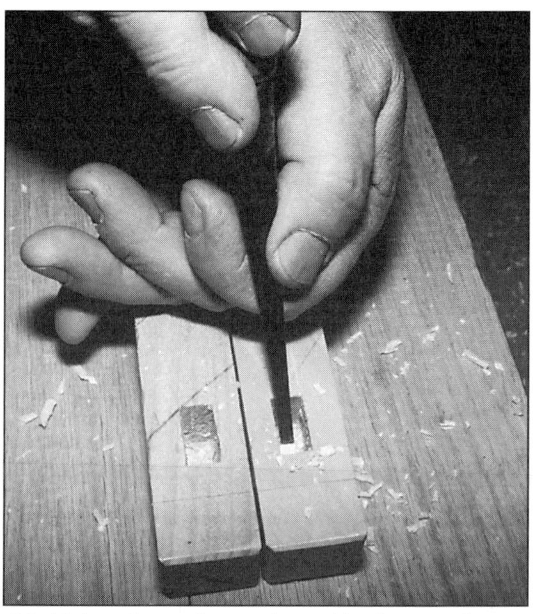

60 Using the *sokozarai nomi* in the mortise at the bottom of the stile.

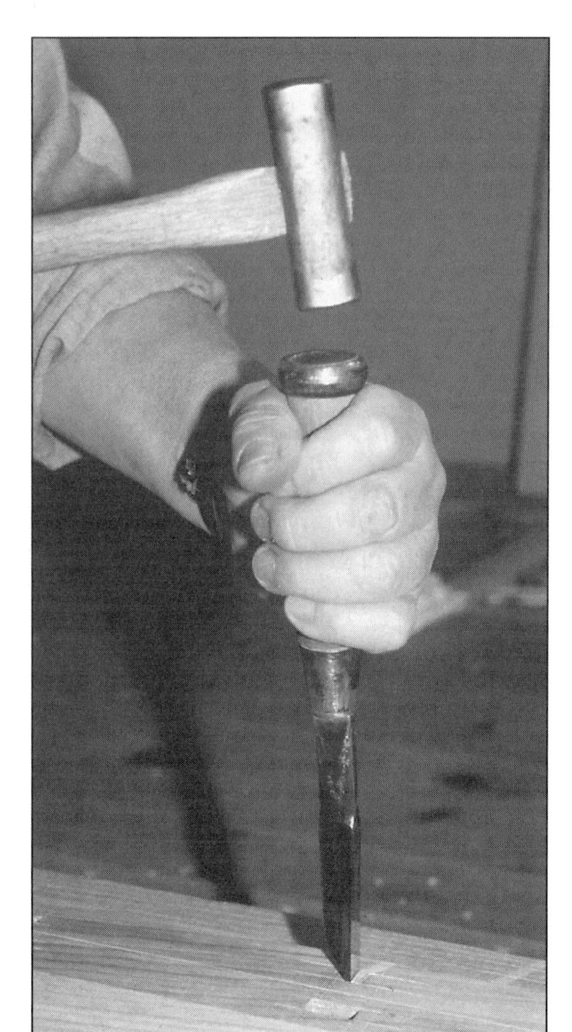

59 Mortising bottom of stiles.

61 For strength, blind mortises must be very deep without breaking through.

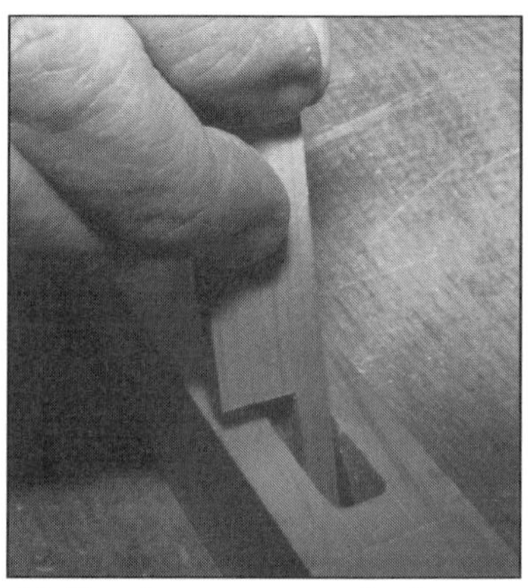

62 Depth gauge made from a *kumiko* scrap indicates when a blind mortise is deep enough.

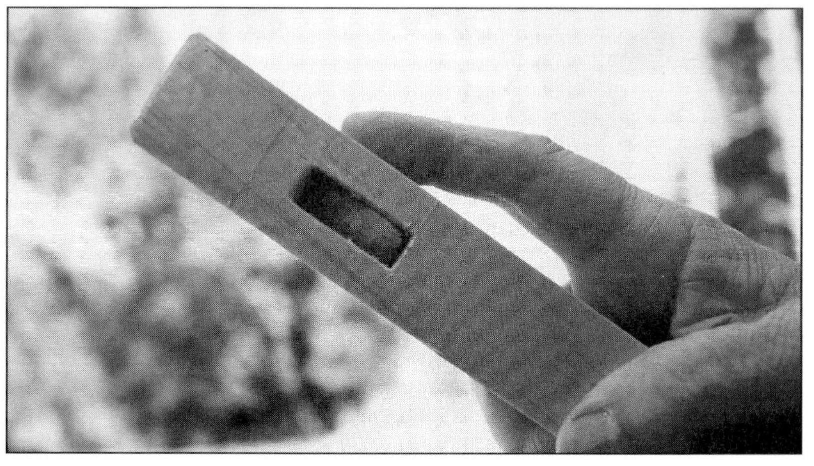

63 One can see light coming through the blind mortise.

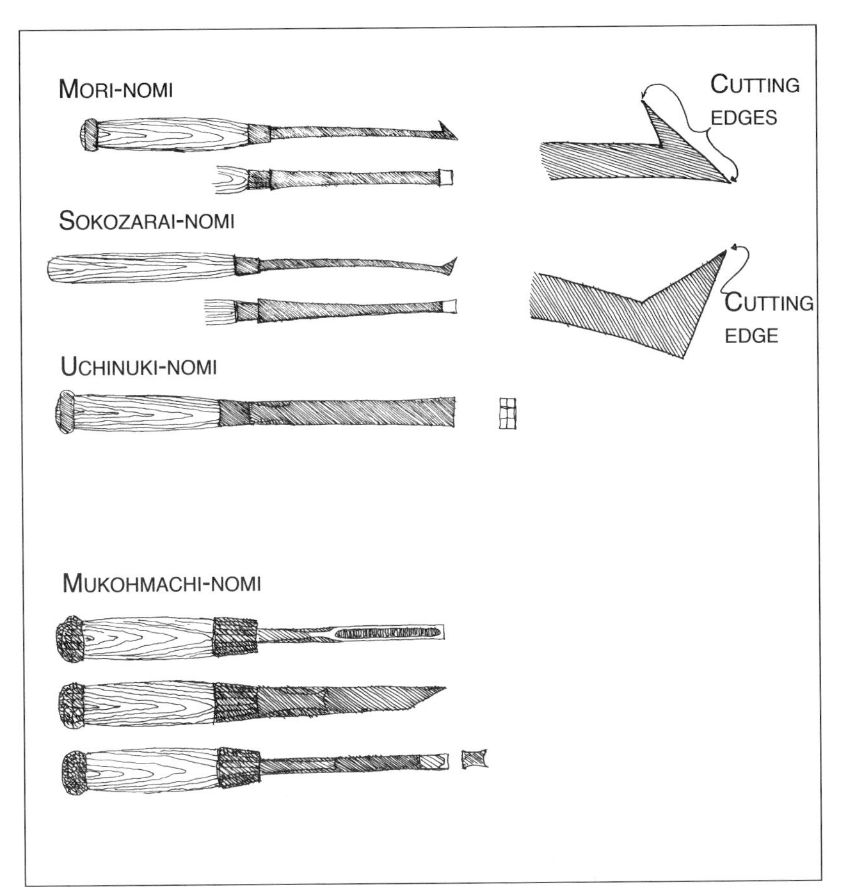

D17 Tools for making mortises.

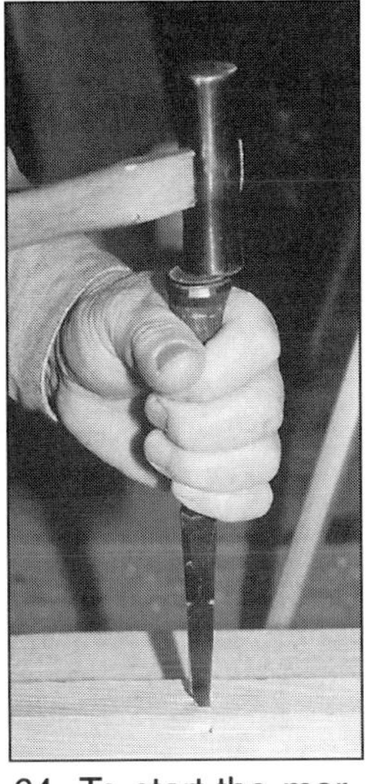

64 To start the mortise, first use mortise chisel.

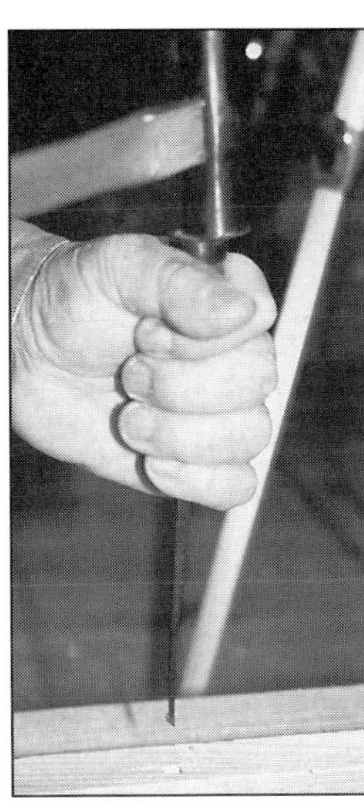

65 Clean out the chips still attached with the *mori nomi* (harpoon chisel) by cutting with a vertical motion.

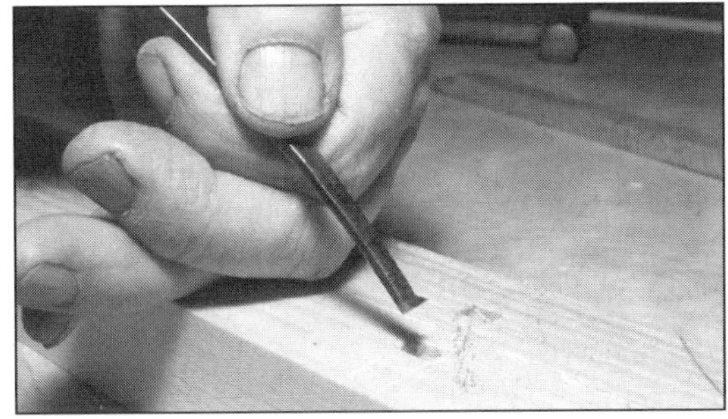

66 Use *sokozarai nomi*, bottom-cleaning chisel. The *kumiko* mortise may be too small to scrape the bottom, but still you can pick up the waste as much as possible.

KUMIKO MORTISES

Kumiko mortises need not be as deep as the mortises for the rail tenons because the subtle bow in the stiles holds the *kumiko* tight. The *kumiko* mortises are a little too small for the bottom to be scraped in the usual way, so you chisel as deeply as possible with a 1/4″ (6mm) mortise chisel, then clean up with a *mori-nomi* and *sokozarai-nomi*. The *mori-nomi* or harpoon chisel cuts on the vertical stroke and is used to lift chips out of the mortise. The *sokozarai-nomi* has a goose neck and a spade-like bend at its end, and is used to scrape the wood at the bottom of the mortise. Take as much waste out as you can without damaging the mouth. Continue cutting to a depth of between 5/16″ (8mm) and 3/8″ (9.5mm), then use a small steel rod called a *uchinuki-nomi*. This tool has a flat end and is the same shape of the mortise but a little smaller (D17). Mark on the *uchinuki-nomi* a depth of about 7/16″ (11mm)—(*kumiko* tenon length is 3/8″ (9.5mm)— and tap it down into the mortise until you reach the mark. This makes the mortise bottom flat and even (photos 64-67).

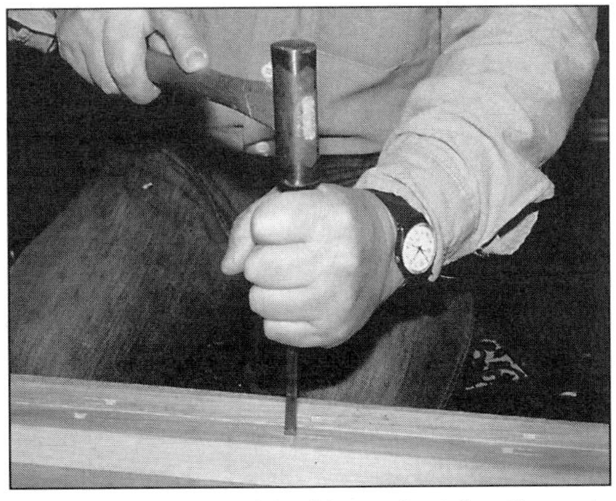

67 Finally use *uchinuki nomi,* strike-through chisel. This tool evens out the mortise depth and clears the bottom. The end of the chisel is scored to keep it from slipping.

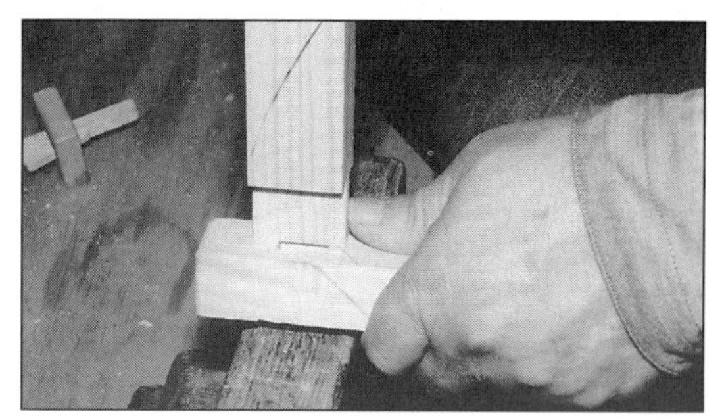

68 I am ready to cut the shoulders of the tenons, but first I check if the tenon's knife marks fit the opening of the mortise.

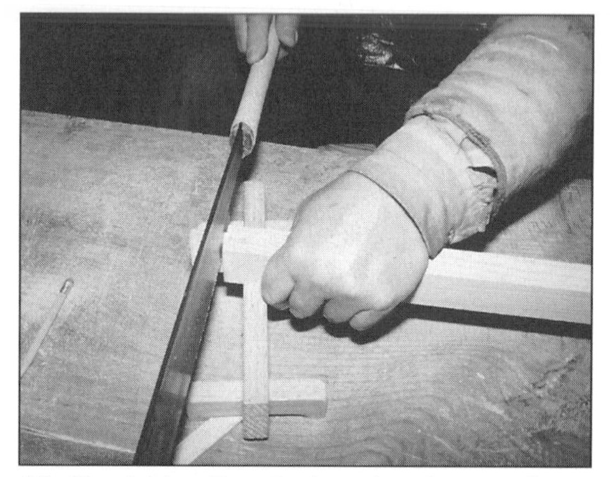

69 By placing the stock on two horses, I can see when to stop the cut of the shoulder. Saw away from shoulders!

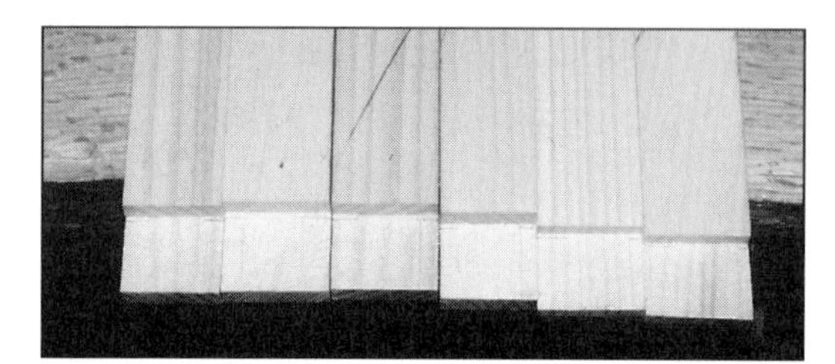

70 Here are ripped tenons.

71 Here I am sawing the tenons' widths.

72 As you know Japanese saws cut on the pull stroke. The long handle is usually held with two hands spaced well apart for maximum power and control. For ripping tenons of rails, I support the stock on one horse so I can see the layout lines on the near edge and on the end grain at the same time. To avoid cutting into the line in or out on either side, saw on an angle into the near edge first, then turn the stock over to cut into the opposite edge, finishing with the saw straight up and down.

SAWING TENONS ON THE RAILS

Tenons on the rails are cut in much the same way as they would be in the West. However, saw the shoulders first, then cut the cheeks (see "The Japanese Mortise-and-Tenon," page 76, and photo 68). To avoid cutting into the shoulder or tenon cheek, you should watch the mark on the nearest edge while at the same time watching the one on the end grain. I saw on an angle, starting at the near edge and continuing toward the end of the end grain. Then I turn the stock over and cut through to finish off. This way the tenon is cut accurately. To cut the narrow third and fourth shoulder, don't saw right on the shoulder line, for the saw set can damage the first two sawn shoulders. Instead, saw a little off from the line and trim with a chisel (photos 69, 72). With all the shoulders cut, saw the length of the tenons 1/16" (1.5mm) shorter than the mortise depth, using the depth gauge that you made earlier for the mortise depth (photos 74-78). Finally, chamfer the ends of the tenons so they will be easy to insert.

73 After cutting the tenon's width, clean with chisel the extra shoulders.

74 Before marking the length of the tenon use a square to make sure the two rails are parallel to each other.

75 Find the length of the tenon with the depth stick used earlier to measure the mortise's depth.

76 Use a square to mark the tenon's length, but make it a little shorter than the mortise's depth.

77 Sawing the tenons to length.

78 Tenon's length is about 1/16" shorter than the mortise's depth stick.

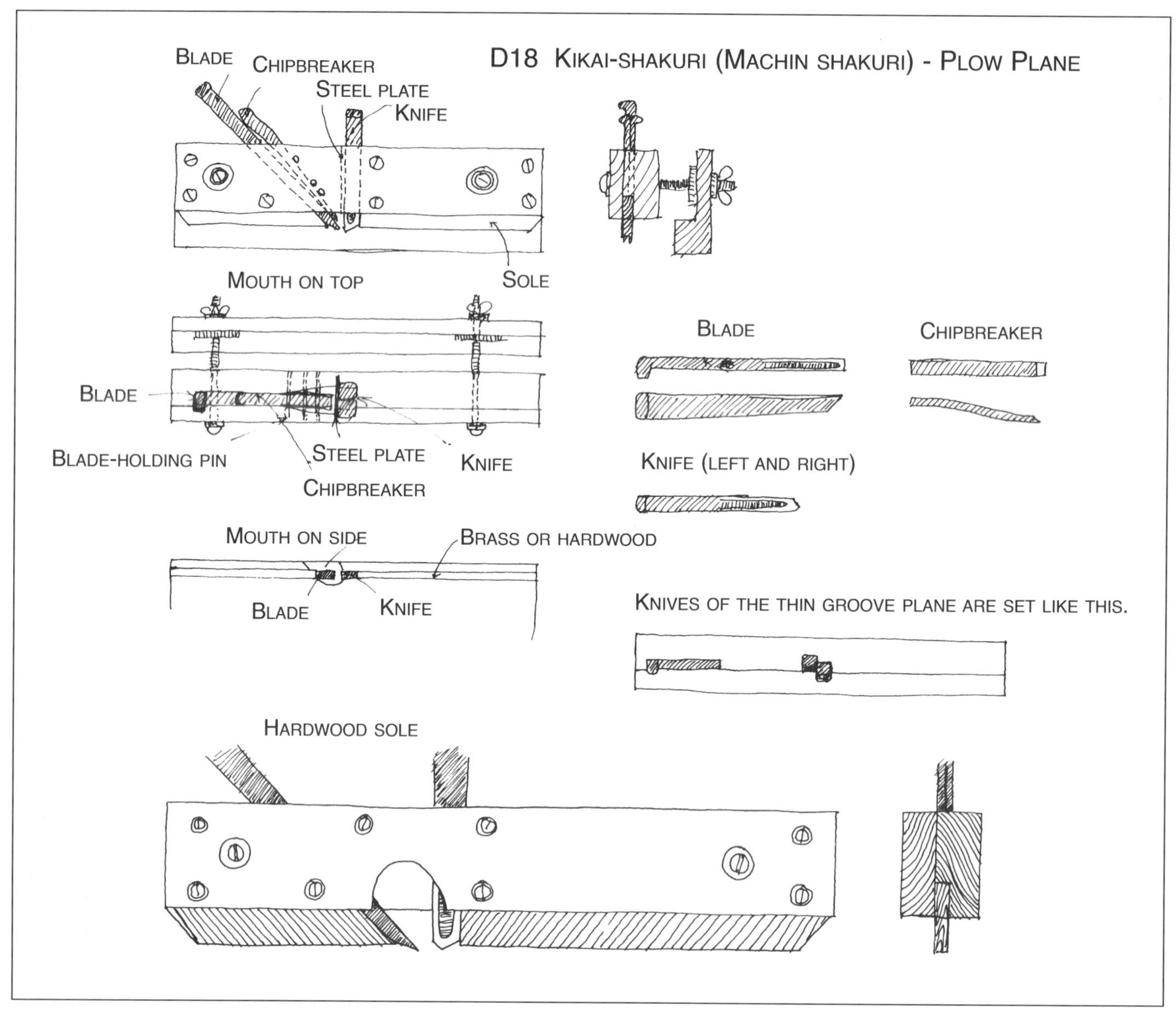

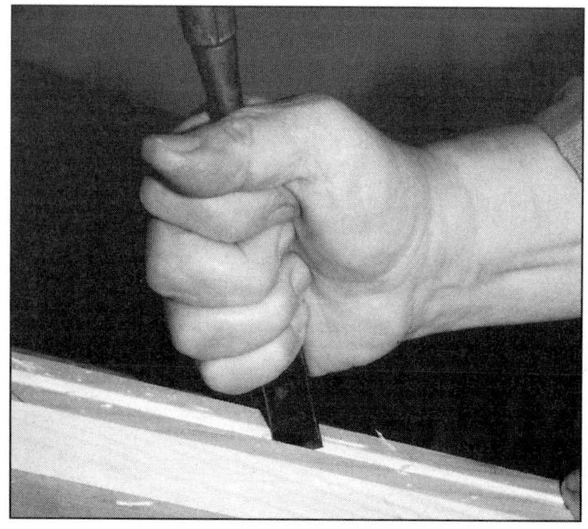
79 Score both edges of groove. Be aware of grain movement. If the grain is curved the chisel might slide out and damage the wood.

80 Finally chisel out wood with mortise chisel.

81 Instead of scoring the stile's hipboard groove with a chisel you can also use the *azebiki* saw.

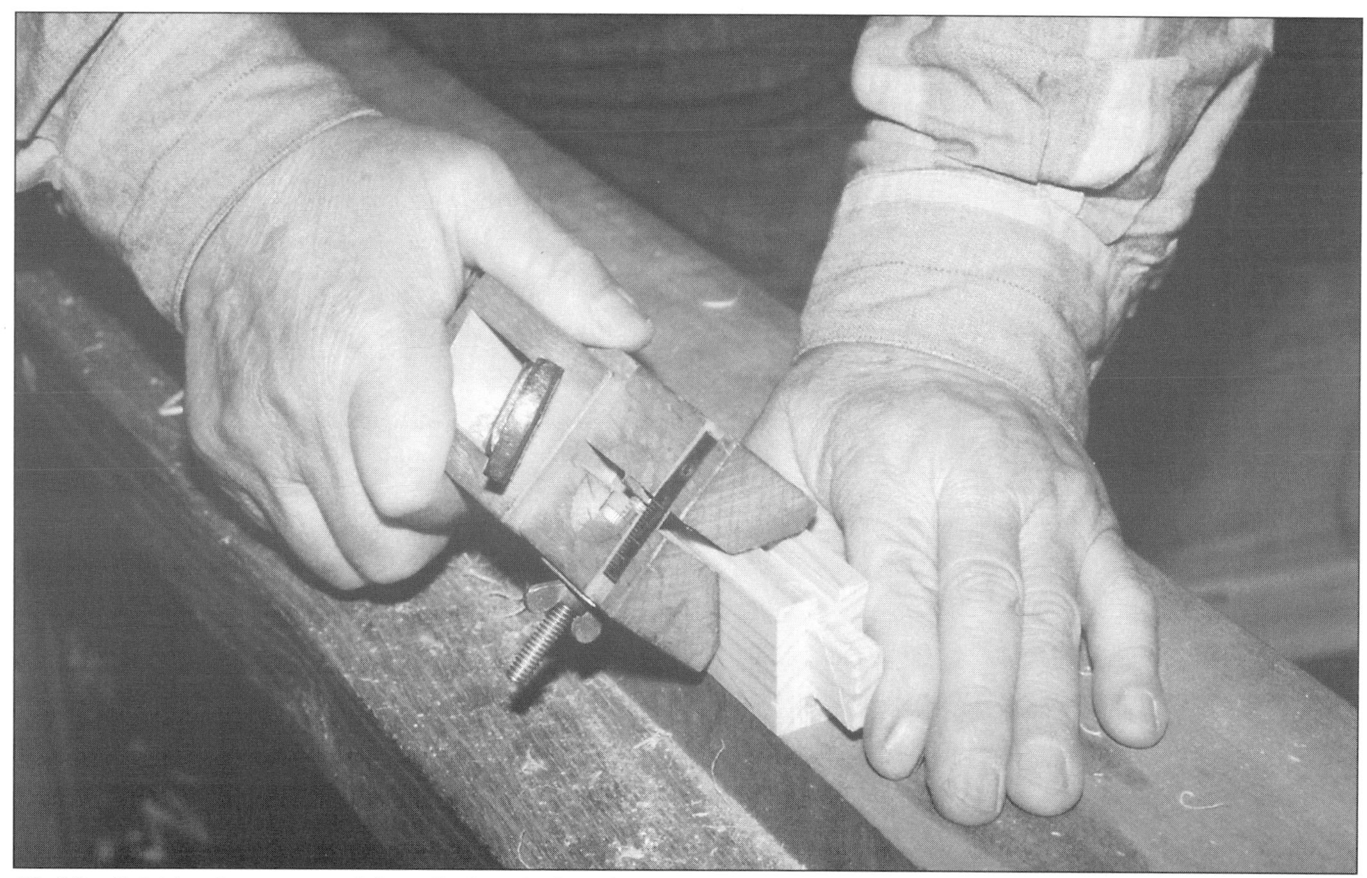

82 After finishing the groove of the rail, chamfer its edges with a chamfering plane.

CUTTING GROOVES FOR HIPBOARD

The next job is to cut the grooves on the stiles—you've already decided the position of the grooves. If you're using a *kikai-shakuri-kanna,* or plow plane, set the fence in the desired position and use it the same way you would a marking gauge (D18). Position it from the front face. You have to be very careful not to damage the other end of the mortise mouth. Because of this plane's structure, you cannot make a square end, so use a 3/4" (19mm) or 1" (25mm) wide chisel to score both sides of the bottom of the groove. Then use a narrower mortise chisel, bevel down, to remove the waste to a depth of about 9/16" (14mm) (photos 79-81). This will allow for hipboard shrinkage. So when you cut the hipboard, the height will be the hipboard opening plus 1/2" (12.5mm) (1/4" top, 1/4" bottom) (6mm), and the width will be the distance between two stiles plus 1" (25mm) (1/2" left 1/2" right) (12mm) plus 1/8" (3mm) for breathing space. When the grooves are complete, you can work on the bottom and middle rails. They are only 1/4" (6mm) through with the length. The next process is cutting the rabbet in the top rail (if the rail is full thickness), and the rabbet in the bottom rail for the track (photo 84).

83 Making hipboard groove on bottom rail.

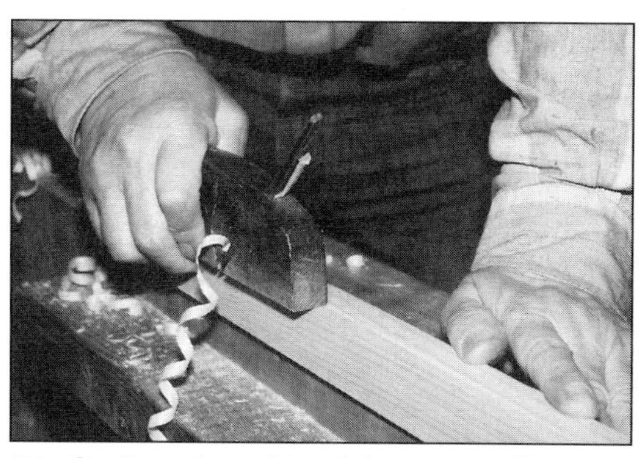

84 On the other side of the grooved bottom rail make a shallow rabbet for the track.

85 When the *tategu-shi* planes his stock he supports the wood on a planing beam, *kezuri-dai*, that is, a beam held at one end by a triangular support and lodged against anything sturdy at the other. A nail driven into the beam stops the work against the pull of the plane. In the countryside during the post-war period, the *kezuri-dai* was fashioned at the work site and left behind when the craftsman finished the job and moved on.

FINISHING THE FRAME

Most of the work is finished except a final planing of all the *shoji* parts, to clean them of handling and layout marks. The Japanese prefer natural surfaces, so the *shoji* receives no further finish. The finishing plane takes off the slightest shaving, with only one or two passes. Pressing the plane hard against the stock burnishes the surface and imparts to the wood a warm glow (photo 85).

After finish-planing each piece, lightly chamfer the edges of the main frame parts, except the inside front edges where the tenon shoulders meet flush with the mortised rail.

86 Now that the frame has been grooved you can mark the exact width of the hipboard.

88 Checking squareness of hipboard.

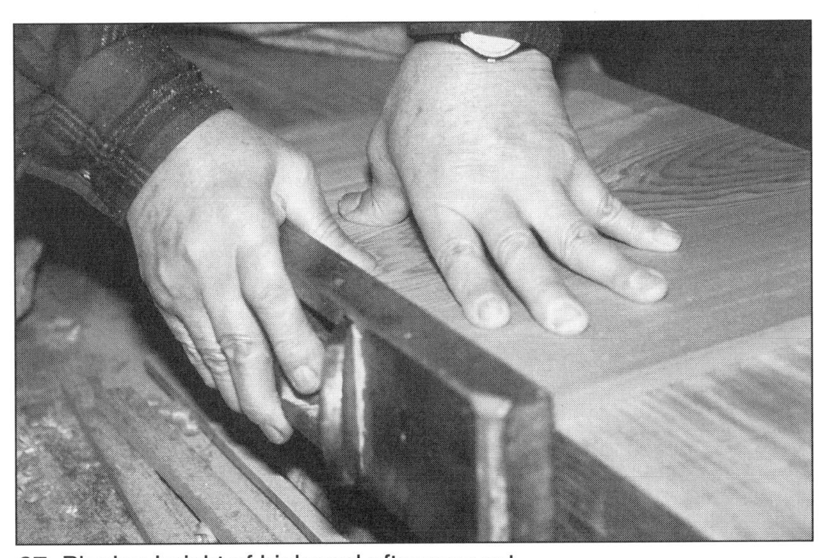

87 Planing height of hipboard after saw cut.

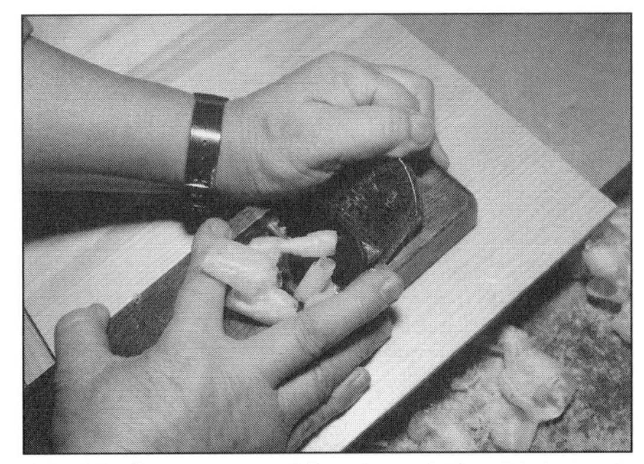

89 Earlier we roughly planed the hipboard. Now use a finishing plane to finish the board with its final shavings.

FINISHING THE HIPBOARD

Cut and plane the hipboard to fit, as previously mentioned. Allow room for the wood to be loose across the grain (the distance from the bottom of one groove to the bottom of the other groove, minus 1/8") (3mm) (D14, page 34), and exactly the same for the height, but it does not need 1/8" (3mm) loose space (measure from the bottom of one groove to the bottom of the other groove). Finish-plane the hipboard's two faces and chamfer all edges (photos 86-90).

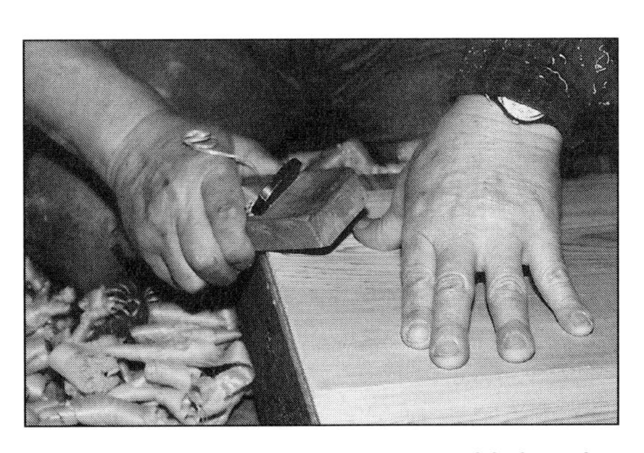

90 For the hipboard to slide smoothly into the groove, finish it by chamfering all edges.

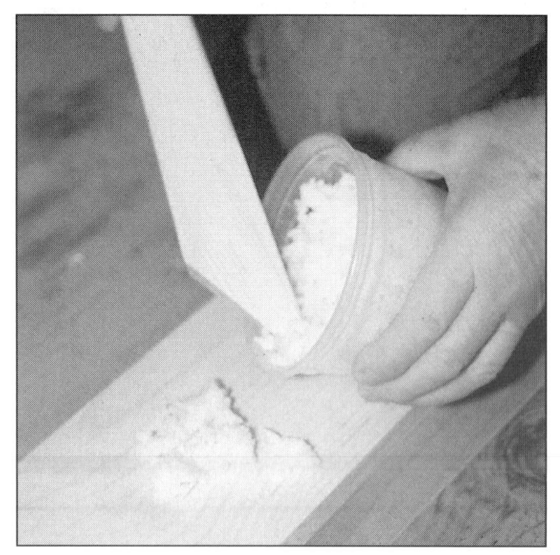

91 To make glue, start with deliciously cooked sticky rice.

92 Mash the rice with the glue stick.

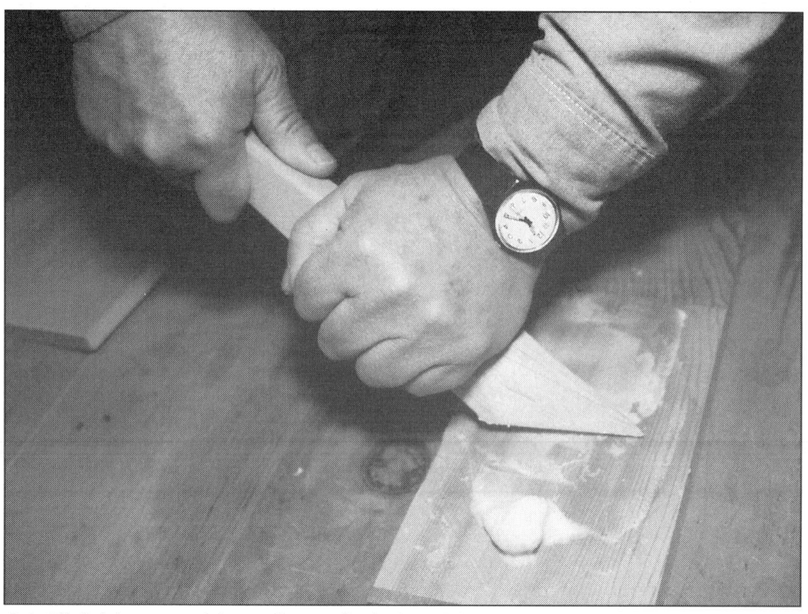

94 Finishing mashing all needed glue. However the glue is still too stiff and hard to be used.

95 Dilute the paste with water to reach the desired consistency.

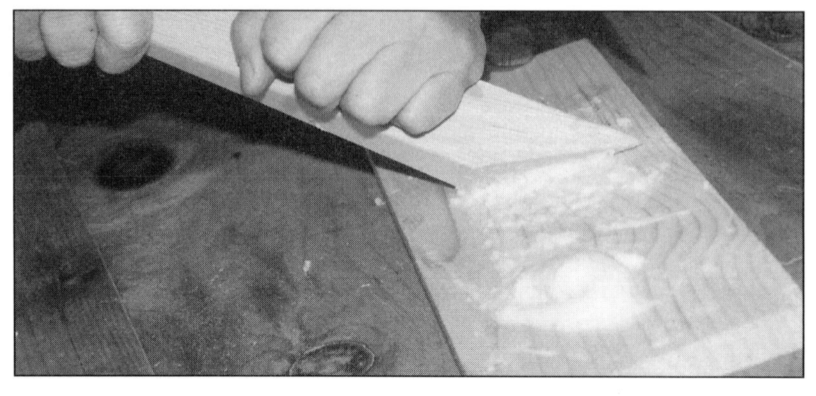

93 Work the rice to an even consistency.

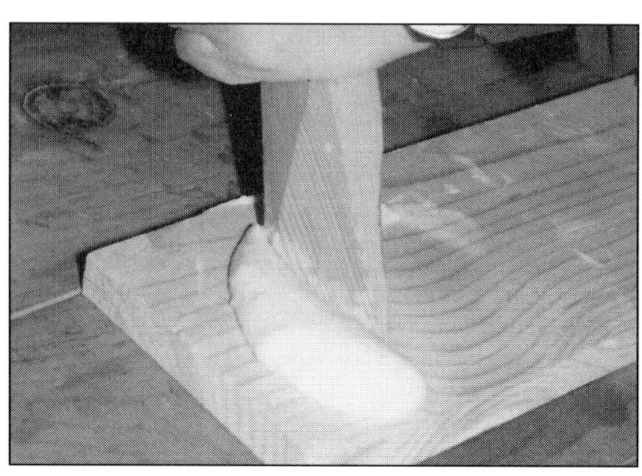

96 Now the desired consistency is found. The glue is ready to be used.

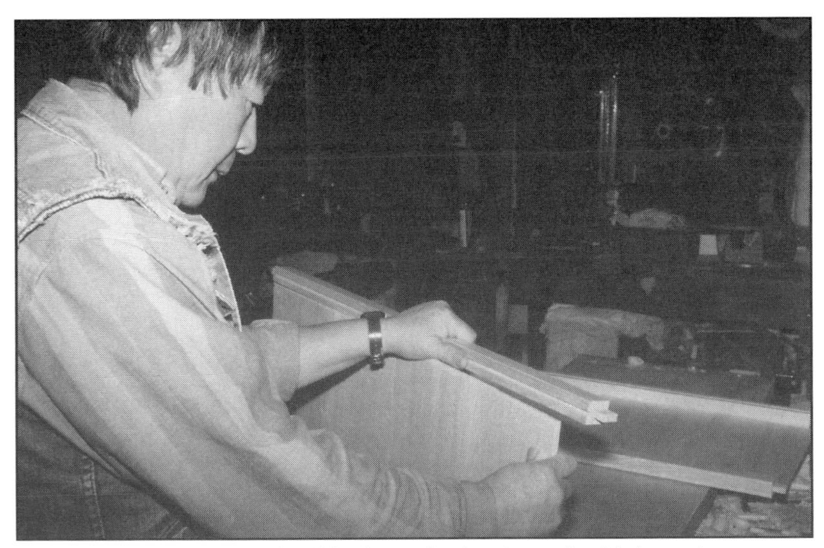

97 Checking groove of rail before placing it on the hipboard.

98 Place rail on hipboard and hammer on a small block of hardwood with chamfered edges to avoid damaging the parts.

99 Two rails are placed on hipboard, notice exposed tenons.

ASSEMBLING THE *HIPBOARD AND RAILS*

Now at last you're ready for assembly. Glue is used for keeping joints together. It simply assists the joint but is also used as a lubricant when assembling, for without glue the sides might scrape. I use rice glue, called *sokui*, that I make myself (photos 91-96). I cook delicious sticky rice, squash the grains with the glue stick, and then dilute the paste with a few drops of water. This glue enhances the strength of the joint without making it impossible to disassemble in case the *shoji* ever needs repair. Any starchy glue, like wallpaper paste, can also be used.

100 (right) Assemble hipboard and rails to *kumiko* lattice. You do not need glue here because the frame joints hold the *shoji* together.

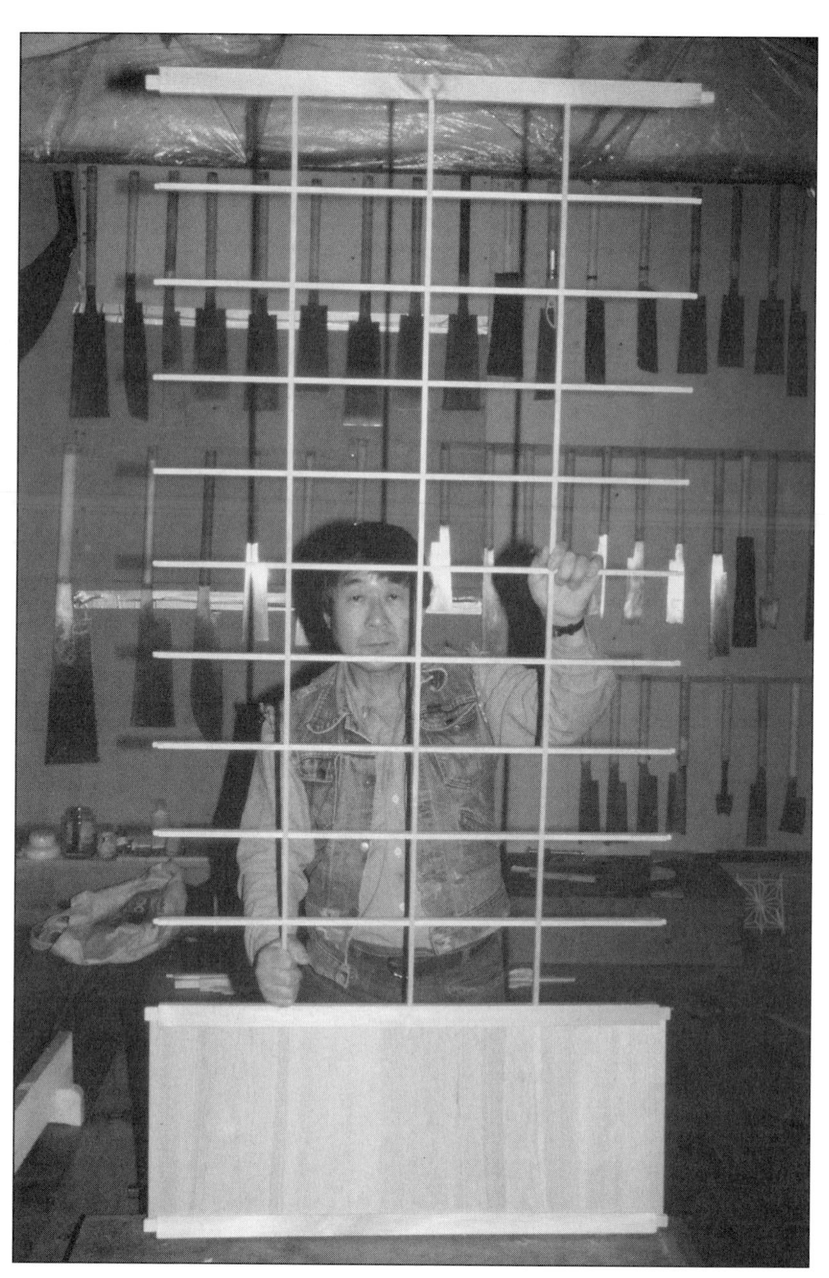

101 Align separately three vertical *kumiko* and nine horizontal *kumiko* together, and put rice glue in the lap joints.

102 Assembling the *kumiko* into their final position. Squeeze horizontal on verticals, tap first lap joint with round-sided hammer, tap third joint, then tap middle. Evenly tap them in. Do not force on one section or you might crack the bottom of the lap cut.

103 As you tap the lap joints hold down the *kumiko* with fingers.

ASSEMBLING *KUMIKO* AND STILES

Assemble the *kumiko* first. Gather the horizontals and verticals in separate groups. Use a skinny piece of *kumiko* scrap to make quick work of applying rice glue to the shoulders of the notches. Don't put glue in the bottom of the notches, because it will keep the *kumiko* from fitting tightly (photo 101). Then tap the *kumiko* together. First lay down three vertical *kumiko*, checking the pencil marks that indicate the bottom of the tenons. Insert horizontal *kumiko* from the top or bottom (photos 102, 103). Fit the *kumiko* assembly into the mortises of the top rail. The assembly is in the same direction as the stiles, so no glue is needed here. Fit the *kumiko* assembly into the mortises of the middle rail.

104 Glue mortises of stiles. Do not glue groove because hipboard breathes and moves, so if glued the hipboard might split.

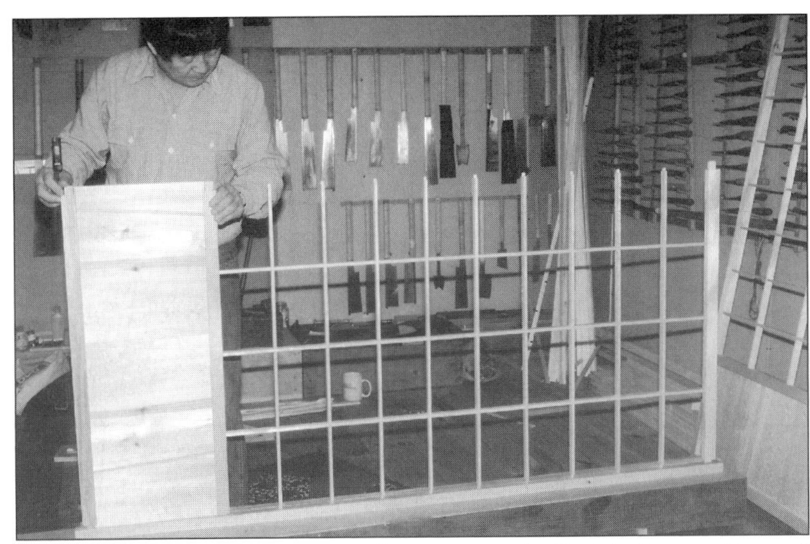

105 After gluing stiles, tap *kumiko* tenon and rails into the mortises of one stile.

106 Before adding the rails, tap around mortise mouth to make sure the rail shoulders will fit.

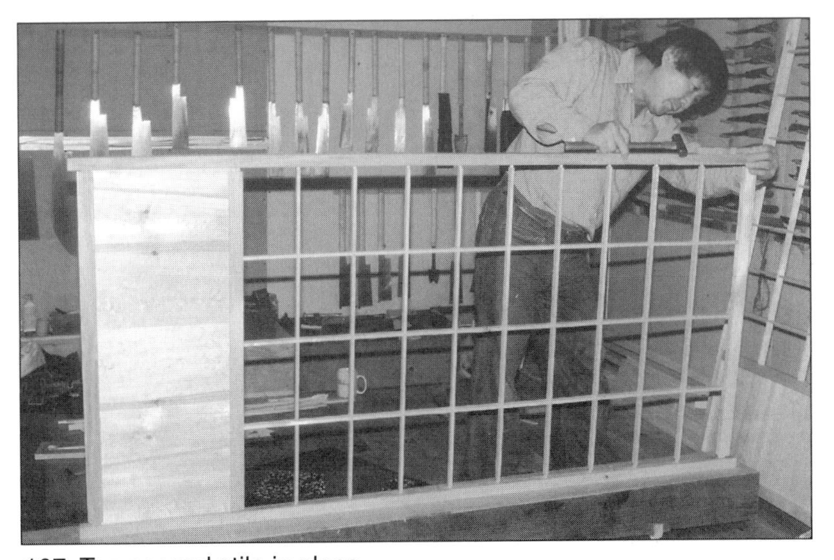

107 Tap second stile in place.

Now you are ready to add the stiles. Position left and right stiles next to each other, inside edges up, then tap around the rail mortises with a hammer so the edges of the rail shoulders will fit tightly (photo 106). Then apply glue to both mortise stiles at once (photo 104). Tap the rail tenons into one of the stiles, stopping when the *kumiko* tenons just begin to engage, then tap the other stile onto the upper tenons with a hammer, using a small block of hardwood with chamfered edges in between to avoid damaging the parts. When the tenons fit tightly, check that the *shoji* is square and free of twist. Tap and twist it into shape if necessary (photos 108-109 on page 48).

108 When all joints are locked into place, check if both stiles are parallel to each other and free from any twist.

109 Also, square stile with rail.

INSTALLATION

With assembly, the tense part of the *tategu-shi*'s challenge is accomplished. Installation is the joy of displaying the work. However, installation is also the moment of truth. A messy installation points to a *shoji* that was not well crafted.

Installation is like icing a cake. You must work carefully and neatly. Place the *shoji* on the outside ledge of the bottom track and check the stile against the door frame or column for alignment. If it fits tightly, cut the bottom horns even with the bottom rail. But if necessary, cut the horns at slightly different heights to align the stile parallel to the door frame or column. Rabbet the bottom of the stiles like the bottom rail to fit the groove in the track, and chamfer all corners of the rabbet. Now put the *shoji* back on the ledge (not in the groove) and press the top of the *shoji* up. Pencil-mark where the inside face of each horn meets the track. This will be the shoulder of the top rabbet. Add 5/8″ (16mm)—this will be the top part of the rabbet. Cut the horn at this mark and rabbet to fit the track, then chamfer all around the rabbet. Now lift the *shoji* and insert the top rabbet into the upper track groove and the bottom rabbet into the bottom track groove. Lastly, check whether the screen slides smoothly.

PHOTO COURTESY MIYA-SHOJI, NEW YORK

110 Common *shoji* divide the rooms and screen the windows and doors.

111 Beautiful light filters through the *shoji* paper.

PHOTO COURTESY MIYA-SHOJI, NEW YORK

PHOTOS COURTESY MIYA-SHOJI, NEW YORK

112-114 *Shoji* may be used in many different ways in a contemporary Western interior. At top left, the *shoji* make an attractive table lamp. At bottom left, the *shoji* have been used conventionally as sliding screens, but also as diffusion screens for overhead lighting.

PHOTOS COURTESY MIYA-SHOJI, NEW YORK

115-117 *Shoji* offer a beautiful and functional way to cover the windows in a Western home (top and left). *Shoji* also make an excellent folding screen in a small apartment.

An Occurrence Under the Overhang

During my apprentice period, I faced hundreds of incidents while preparing materials. One incident I remember very clearly: the sound, the view, the pain, the smell, the taste, and the loud voice.

It was July, one hot summer afternoon. My master and I were working at home, and my mother was in the house. We were making *shoji* under the overhang in front of the house. I was then planing the stiles on a planing beam and my master was sitting on a mattress about ten feet away from me. I had already finished two stiles and the third one I had just given to him. As I went back to the planing beam to start on the fourth one he shouted at me, "Toshio, come here." "What is it now?" I thought to myself. "Take a look at this," he said and pushed the end of the stile in front of my eyes. Obviously he felt disgusted by the way I had planed the stiles. I jerked my head backwards and closed one eye to look at it. I wasn't really looking at anything, instead I said within my heart, "What's the big deal." I am sure that my face and body were speaking out very clearly and loudly. I grabbed the stile and turned to go toward the planing beam, but at that moment he snatched the stile from me. I turned my head around, then I saw the stile straight up in the air with his two fists holding tightly on one end. Everything happened so fast. Spontaneously, I tucked in my head and covered it with my

hand. The next moment I felt a harsh shock, and heat with pain followed on the right side of my back. I held the middle of my shoulder blade while he was shouting something or maybe I was screaming so loudly. I ran to the other side of the house, sat on the ground and leaned against the wall. My eyes filled with tears as the pain was doubling. I removed my left hand from my back and looked at the palm. Was I expecting blood? I did not know. There were two wet and reddish broken blisters, which were made the day before from handling a new tool. I started to pick them with a chip of wood. I did not know why. Perhaps I was trying to ease the pain from my back.

My mother could not interfere on any matter between me and my master. This was an unwritten rule of most apprenticeships. It must have been very hard for her to watch her son being beaten by his stepfather (my master), however, she swallowed her tears and kept herself away from the matters between master and apprentice.

She knew what had happened, and she came to me—not between us. I continued poking the wound on my palm, now even harder and deeper, again for no conscious reason. Maybe I wanted to receive sympathy, tenderness from her, or care and concern. "Stop that," she said, as she took the chip and pushed my hand away from the wound. I was sobbing, "I can't, I can't continue." I touched the pain with my blis-

tered hand. It was very hot. " I want to be a truck driver." (My master's younger brother was a very nice man and he drove a truck all over the country. He often told me of the roads he traveled and the places he went to.) "No," she said gently, "be patient, there is an old saying that says when one achieves a skill in one's heart one never starves in one's life. Once you learn the skill then nobody can take it from you, this skill can help your life. Be patient," she said, "be humble and obey him. Learn everything he has." She paused a moment, and all I could hear now was the sound of my sobbing. She continued, "Well, wash your face. Straighten yourself up then go to him and apologize." Silence followed. She then went into the house. I gently touched my shoulder again, it was swollen up with the mark of the stile. I stood up and walked through the working space, behind my master, into the kichen. My mother had a tender smile on her face. Without a word she drew a wooden bucket from the well and filled the washbowl with fresh cold water. I washed my face, and she handed me a towel. Then, I went to my master and apologized for my behavior. There was no reply, however, I knew that my apology was accepted. I went back to the planing beam and tried to resume planing the stile. Suddenly tears filled my eyes again and I couldn't see the planing surface. I was seventeen years old.

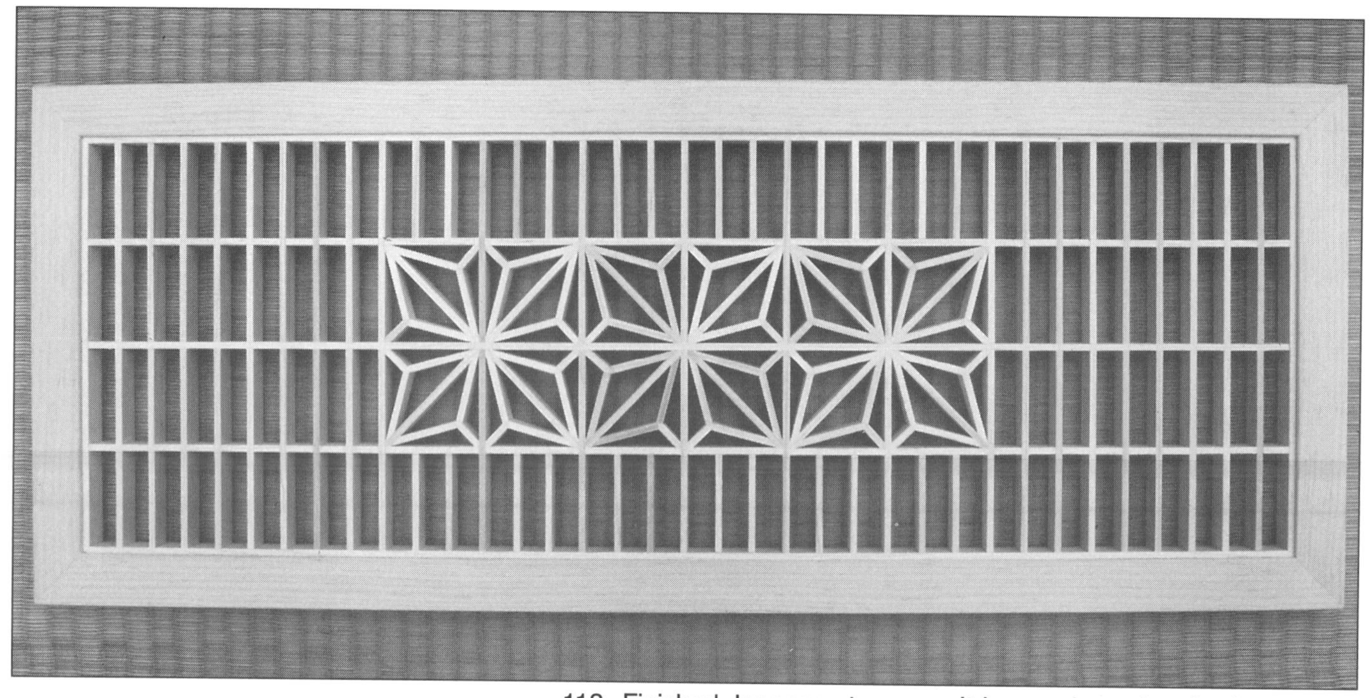

118 Finished Japanese transom. It has an inner *kumiko* frame containing the *kumiko* and the *asanoha,* or hemp leaf pattern, plus an outer frame. Some of the *kumiko* tenons extend beyond the inner frame, into mortises in the outer frame. Because of this, the outer frame must be assembled around the *kumiko* frame. All the corner joints are mitered. The outer frame uses a strong boxed mortise joint. This is a very special project that will teach you much about traditional Japanese woodworking techniques and craftsmen's attitudes.

The Japanese Transom

This type of transom is made without paper or glass and is usually found between two rooms, inside of a veranda, and above sliding doors. As with the common *shoji,* this project illustrates much of the traditional Japanese woodworker's attitude, philosophy, and aesthetic values. From time to time as I describe the making of the transom, I will point out unique aspects of Japanese woodworking.

Start by making a full-size drawing. This transom has a mitered outer frame and a mitered *kumiko* frame. Its main pattern is called *asanoha,* or hemp leaf, which is one of the most popular traditional Japanese designs.

Prepare your materials according to the drawing. Make sure to allow for more length than called for by the actual drawing measurements. Squareness and accuracy are essential.

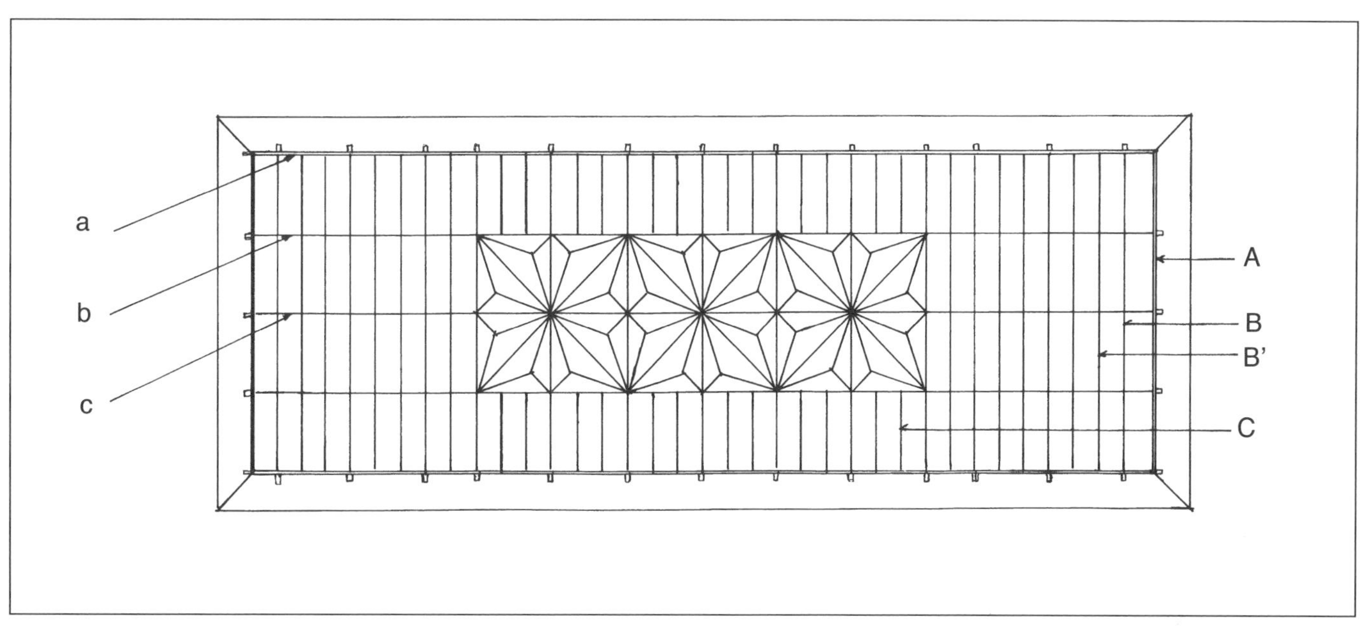

D19 The transom with *asanoha* leaf pattern requires three different horizontal *kumiko* (a, b, c) and four different vertical *kumiko* (A, B, B', C).

119 Layout of frame with assembled *kumiko*. Small tenons lock *kumiko* into the frame.

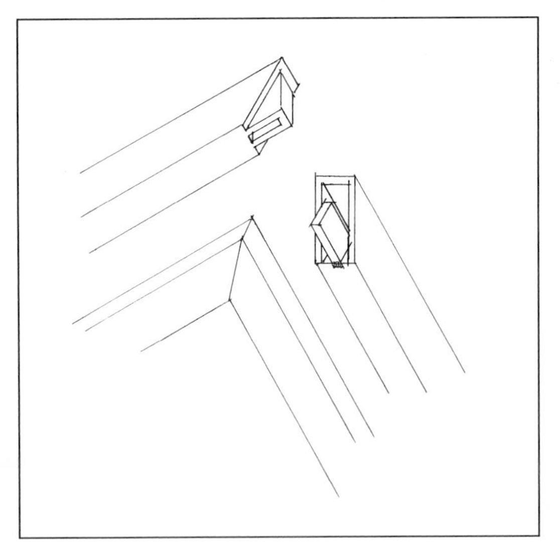

D20 Mitered and boxed mortise-and-tenon joint.

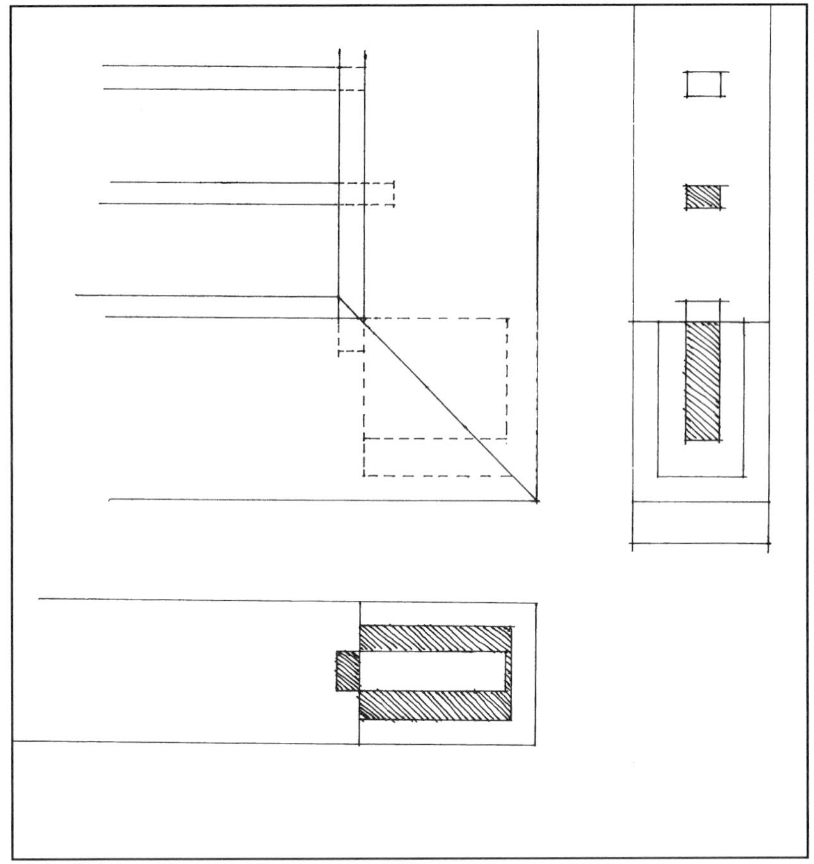

D21 Mitered and boxed mortise-and-tenon joint.

MARKING THE FRAME

First is marking and cutting. Mark the frame first (D23). Commonly, sliding screens and panels have mortises on the vertical pieces (stiles), and I made this project that way. However, I suggest you make the mortise on the horizontal pieces for easier assembly, and I have made the drawings that way.

The corners of this project are mitered and the joint hidden. This special hidden joint is usually used when corners are exposed. Also, this joint is very strong, since each section has, theoretically, both a mortise and a tenon (D20-21). Mark the corner joint and all *kumiko* positions on the frame, even though some of these marks won't be cut into the mortises. After marking all the lines, indicate which *kumiko* line will be mortised. For this particular project, I add about 1/8″ (3mm) to both ends of the frame and cut off the line to simplify marking and cutting.

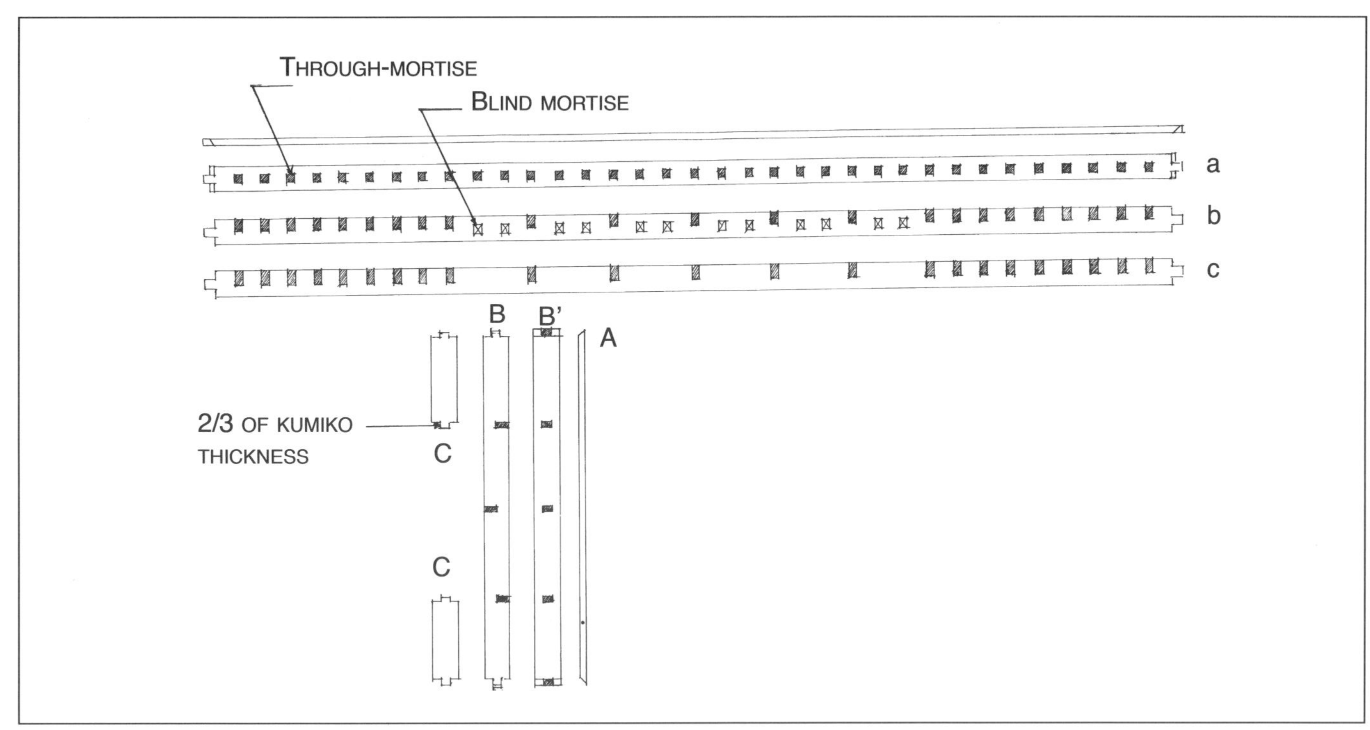

D22 *Kumiko* layout for transom with *asanoha* pattern.

It is difficult to evenly mark and measure all the lines and spaces, so I will give you a few tips. You will mark *kumiko* lines and spaces on a small stick of *kumiko* material, just as you would do when making a common *shoji* (D23). But first you will calculate the spacing of both stile and rail on paper. Unlike the common *shoji*, you don't have to concern yourself with the size of the *shoji* paper.

In this project the vertical frame has only four spaces, so I will mark one space and two *kumiko* marks on the stick. Since the horizontal frame will be mortised at every third *kumiko*, make four *kumiko* marks on the stick with a marking knife (you can do five marks if you wish). Accurate and precise marking is easier to achieve when you use a double marking knife. This very important and convenient tool is especially useful when marking numerous narrow lines. Like any other *kumiko* marking, I suggest you adjust the double knife a hair narrower than the actual *kumiko* thickness. Even after cleaning and planing a little off the finished *kumiko*, keep in mind that the tenon and mortise or lap joints should still be snug.

After marking one section on the stick, lightly transfer the marks onto the frame. However, as you repeat these markings, they will not always match the last line, so you will have to repeat and readjust. It is important that all your marking spaces be as even as possible. You will find that this way of proceeding is fast and accurate.

The next step is marking the *kumiko* (D22). This project calls for many different kinds of *kumiko* cuttings.

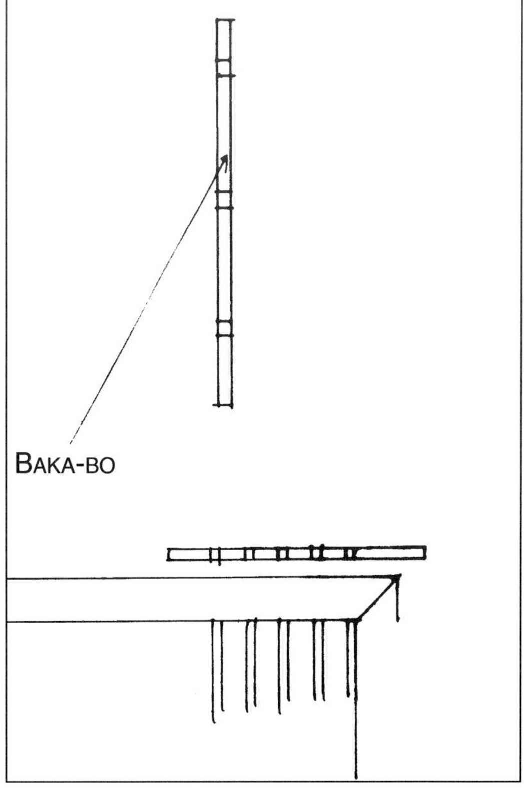

D23 Make a *baka-bo* (marking stick) that spans four or five *kumiko*.

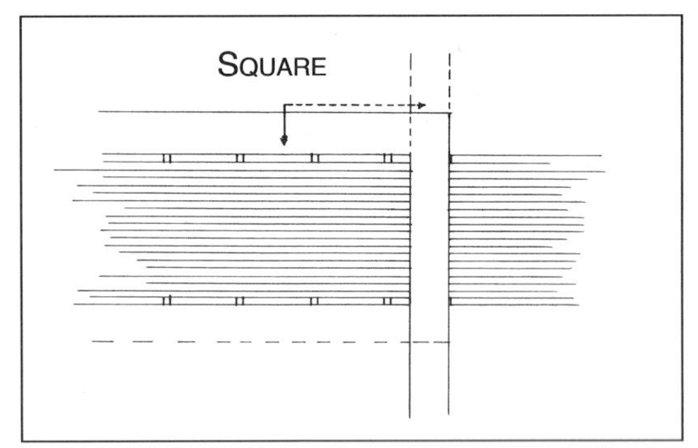

D24 Use square to align markings of the two end *kumiko*, to transfer the marks onto the unmarked stock.

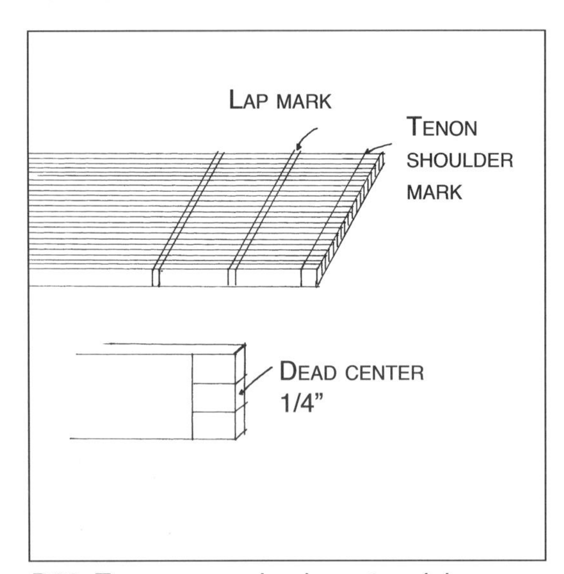

D26 Tenons are dead centered because when assembling the *kumiko* the faces will be alternated. This way the position of the tenons does not change.

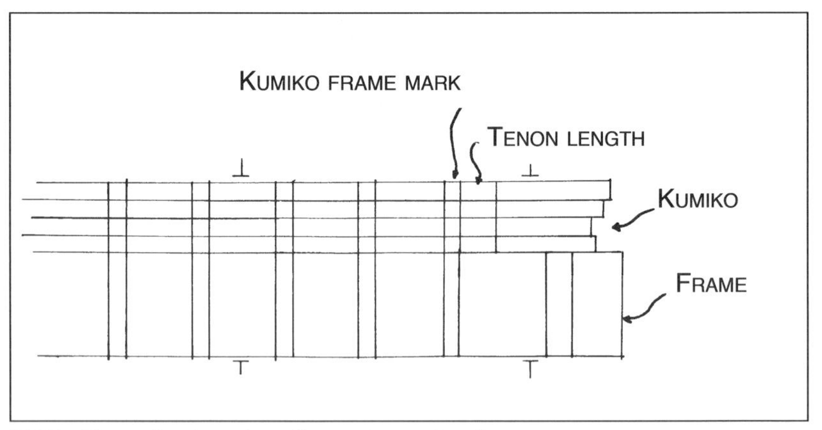

D25 Transferring *kumiko* cut marks from frame onto four *kumiko*.

MARKING AND CUTTING VERTICAL *KUMIKO*

A = vertical *kumiko* frame:	2 pieces
Long vertical *kumiko*:	23 pieces
B = long tenon *kumiko*:	13 pieces +1 extra
B′= short tenon *kumiko*:	10 pieces
C = short vertical *kumiko*:	12 pair. Will be above and below leaves.

Mark and cut the vertical *kumiko* first. Success in marking and cutting depends on the accuracy of repeated measurements. The key is to mark and cut most pieces together. For instance, *kumiko* A, B, B′, and C all have the same tenon shoulder line, as well as the inside of the *kumiko* frame miter line. Thus all vertical *kumiko*, including the *kumiko* frame, could be marked together, but for this particular project I mark four pieces from the frame (D25). Two of the four *kumiko* will be frames that won't have tenons or half-lap cuts, yet you still need all the *kumiko* marks on them. Once marked, put the two *kumiko* frames aside and sandwich all other 34 *kumiko* between the remaining two *kumiko*. As mentioned before, make sure all is square. Transfer the *kumiko* marks (lap cut marks) and the tenon shoulder marks all around (D24). You should cut the half-lap alternately on vertical *kumiko* to minimize the distortion after assembling the screen. Don't cut the 12 short pieces, situated on the top and bottom of the leaf, to size for a little while. Next, mark the length of the tenons—3/8" (9mm)—on both ends and cut them squarely. Then use a 1/4" (6mm) mortise marking gauge to mark the tenons as described in the section on common *shoji* (D26).

Cut the lap joints to uniform depth next. All 36 *kumiko* are still clamped together. Cut a small piece of *kumiko* material to cover 24 *kumiko*, plus 1/2" (12mm). Cut another piece of *kumiko* material to cover the remaining 12 *kumiko*, and again add 1/2" (12mm). Remember that one

120 Tenons and center lap cut with two securing pieces on verticals.

121 Unclamped and separated, two different types of *kumiko*, one for long tenons, the other for short tenons.

kumiko is extra (photo 120).

Usually we cut all laps or mortises before we cut the tenons. However, in this project you can't cut the other two lap joints together. The short ones won't have lap joints—instead, they will have a tenon for the blind mortise. I thus cut all tenons on both sides before cutting the two lap joints. You can cut the shoulder with the saw and remove the waste with a paring chisel. After finishing the tenon on one end, put a pencil mark on its cheek. Now release all clamps and part the long pieces to be and the short pieces to be (photo 121). Then reclamp the long pieces (24 pieces) and finish off the two uncut lap cuts. Remember, the cuts will be on the reverse side of the first center cut because vertical pieces have alternate lap joints (photo 122). Make another short *kumiko* material and tap it into the new groove to cover 10 pieces. These pieces have short tenons (one *kumiko* thickness). Remove the clamps, then take out the piece in the center groove—now 14 pieces are free. Chamfer all four corners of the tenons. Now the long pieces with 3/8" (9mm) long tenons are finished; 10 long pieces are still together. Clamp them and mark the tenon length, one *kumiko* thickness on both ends. Cut them and take out the piece from the groove and chamfer the tenons. Check the pencil mark on the cheek, if you cut it off remark the same side. Now all 24 long pieces are finished.

Next are the short pieces (top and bottom of leaf) (photo 123). Reclamp the 12 pieces, and cut on both sides the length of the tenons one *kumiko* thickness. Check the pencil lines. Next, mark all around the one lap cut mark closer to the tenon. This will be the tenon shoulder for the blind mortise. Cut this tenon length two-thirds of the *kumiko* thickness. Mark the tenon like the others and cut. Remove the clamps and chamfer the tenons. Now all the vertical pieces are finished.

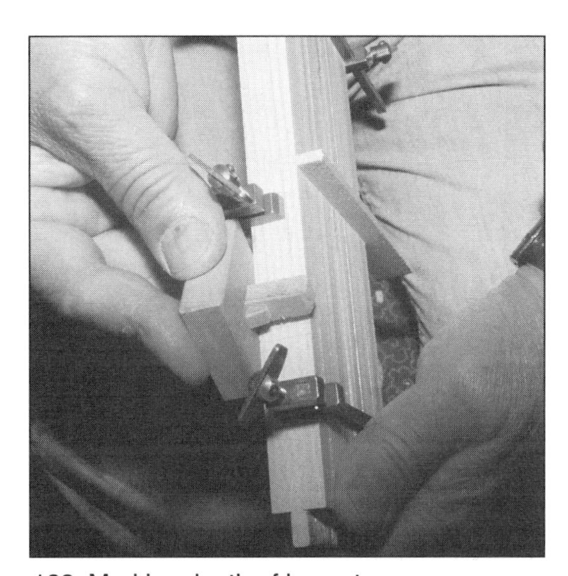
122 Marking depth of lap cut.

123 Finished tenon of short pieces.

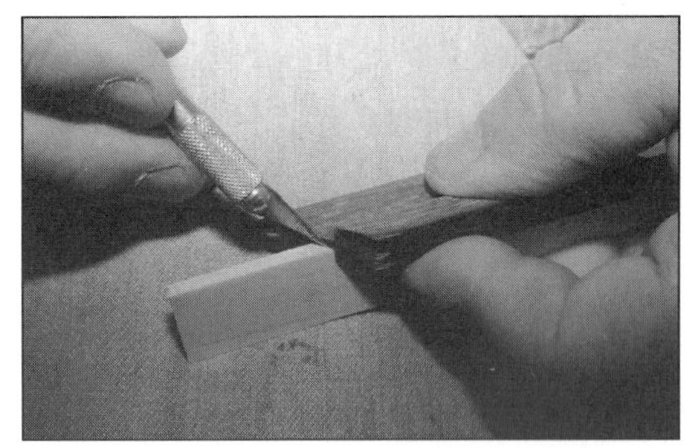

124 Marking miter line on *kumiko* frame.

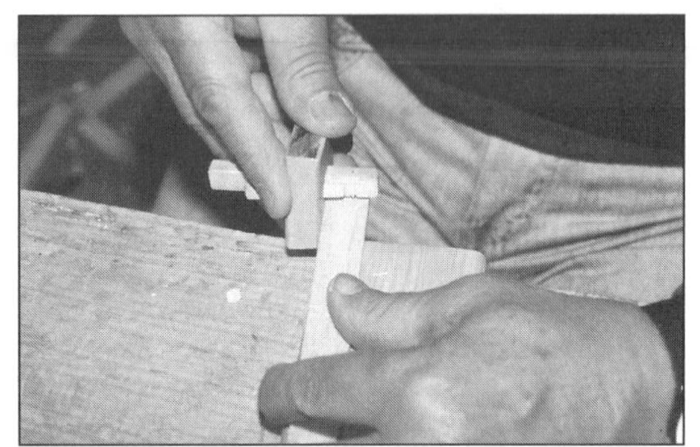

125 Marking mortise on miter corner of *kumiko* frame.

126 Mortising vertical *kumiko* frame.

127 Cleaning mortise on *kumiko* frame.

128 Finished vertical *kumiko* frame.

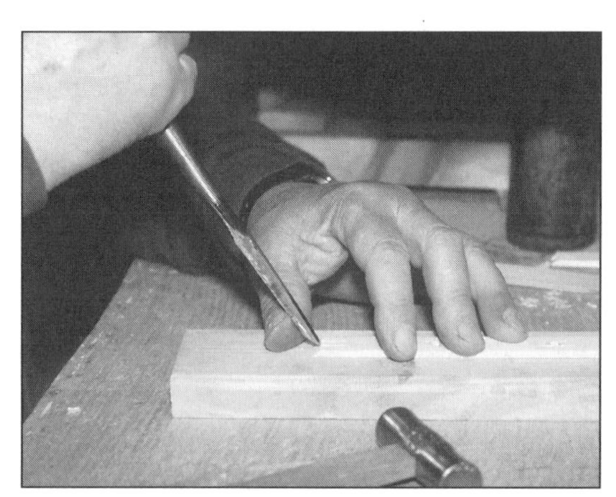

129 Mortising miter corner of *kumiko* frame.

VERTICAL *KUMIKO* FRAME

Next is the vertical *kumiko* frame. Both corners will be mitered and mortised. Both ends of the *kumiko* marks, two lines, will become mitered joints. Keep the ends of the *kumiko* about 1/16″ (1.5mm) longer for easier marking and cutting of the miter joint. Wrap these two lines all around, and on both width sides mark the other three *kumiko* marks. These will be through mortises. Now use the miter gauge to mark the miter lines on both sides (front and back) and mark the mortise lines on both width sides (photos 124-125, D27). First, finish the three through mortises. Cut from both sides (photo 126-127). Then, cut this miter with a paring chisel and miter jig. Cut out the center 1/4″ (6mm) with the mortise chisel (photo 129). Now you've finished the vertical *kumiko* frame (photo 128).

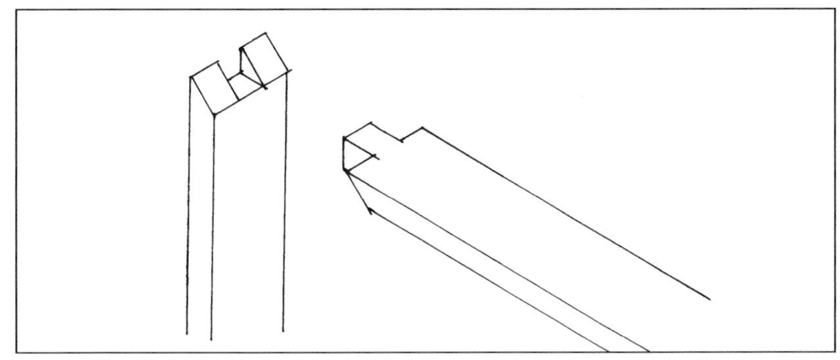
D27 Mitered corner of *kumiko* frame.

130 Marking miter on horizontal *kumiko* frame.

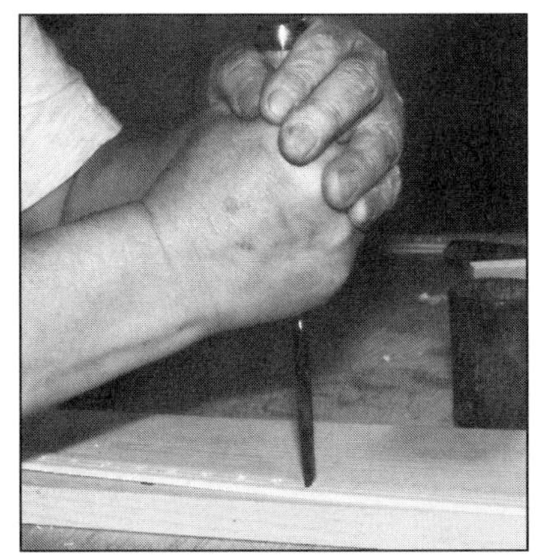
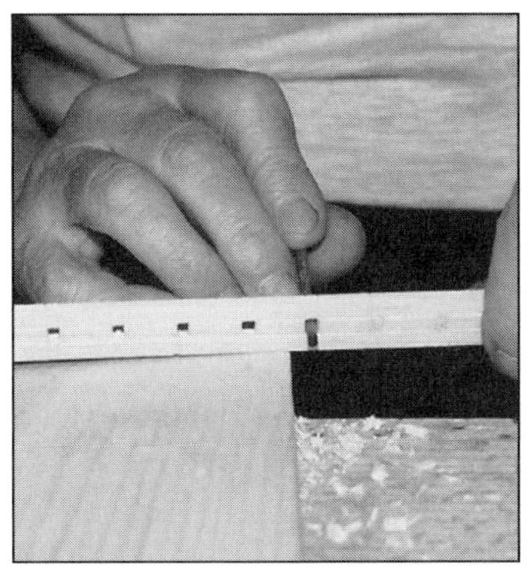
131-132 Mortising horizontal *kumiko* frame.

133 Cutting miter of horizontal *kumiko* frame.

MARKING AND CUTTING HORIZONTAL *KUMIKO*

a = horizontal kumiko frame, 2 pieces
b = horizontal kumiko with lap cut and blind mortise, 2 pieces
c = horizontal kumiko with lap cut, 1 piece

Next make the five horizontal *kumiko*. However, I suggest making one extra piece in case of damage while working. Unlike the vertical pieces, I clamp the six horizontal pieces to the rail and transfer the marks. Do as I explained in the chapter *The Common Shoji*, page 14. Then cut the 3/8" (9mm) long tenon. Lift two *kumiko* from the others with the procedure mentioned in *The Common Shoji*, while leaving one clamp. Clamp the four pieces and remove the last clamp. The two removed pieces will become the *kumiko* frame; wrap the frame layout lines all around. Also mark the other *kumiko* marks on both width sides. Use the same mortise-marking gauge to mark the mortise on the *kumiko* frame and the tenon of the miter joint. Next, cut all the through mortises, then use a paring chisel to finish the miter and tenon. Next is the horizontal *kumiko*. The inside of the last *kumiko* mark will be the tenon shoulder, so wrap the line all around. Also transfer all the other marks to both width sides. Then mark the tenon with the mortise-marking gauge.

134 Testing miter corner joint.

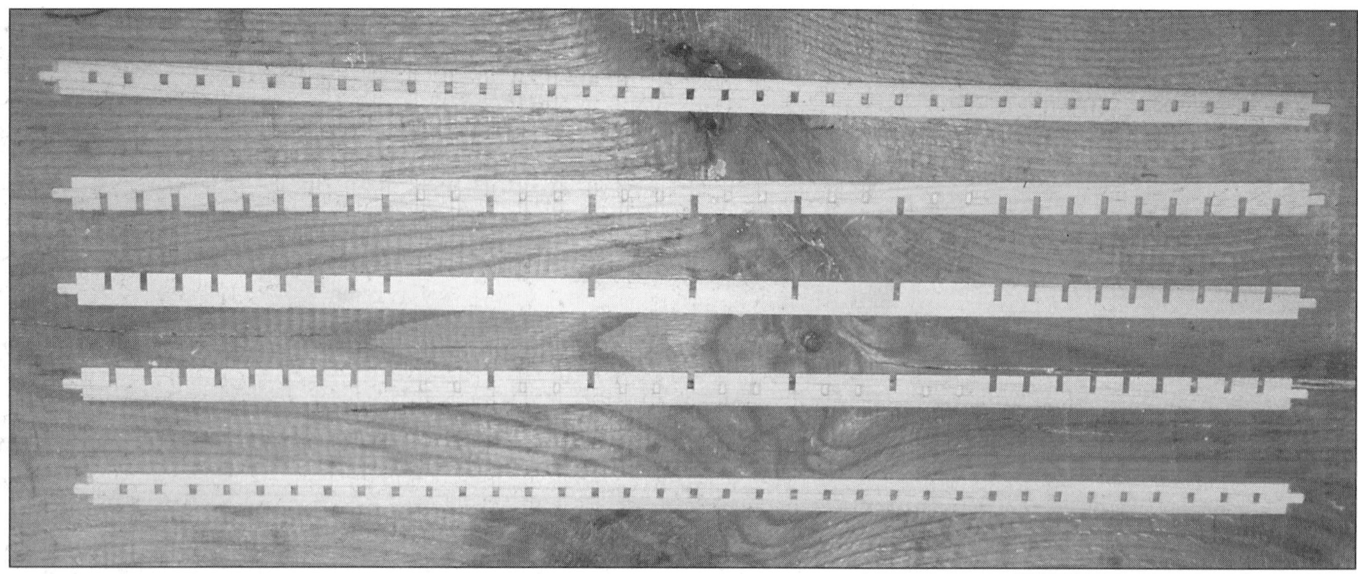
135 Finished horizontal *kumiko*.

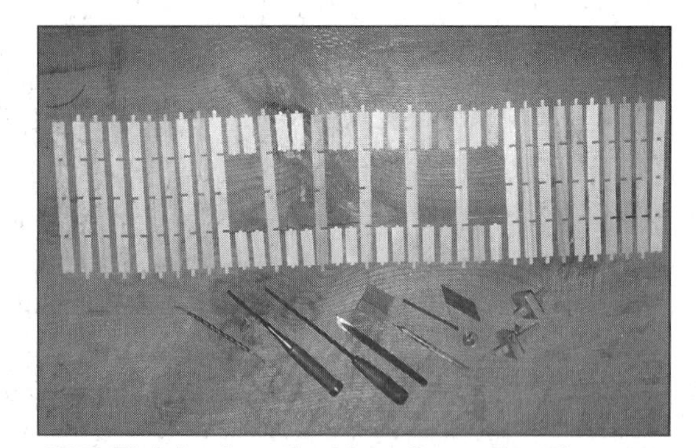
136 Finished vertical *kumiko* and tools.

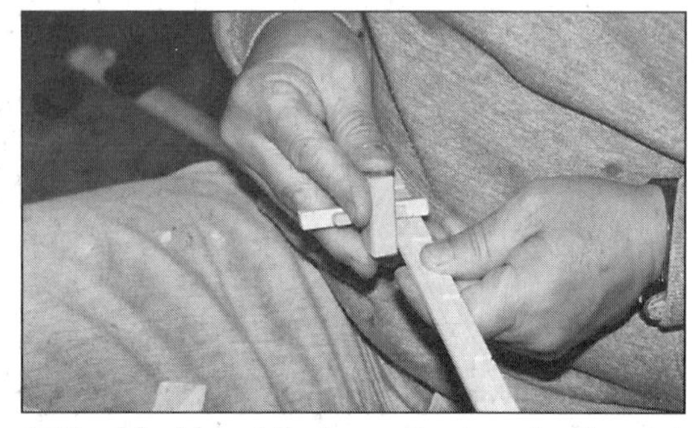
137 Marking blind mortise on horizontal *kumiko*.

The *shoji* maker usually uses a weaving method to assemble the *kumiko*. In this project, there are so many vertical pieces and they're so close to each other that it becomes quite difficult to assemble them with the common method. Instead, they're often assembled with straight lap joints. All horizontal *kumiko* lap joints are cut from one side (D22, page 59, b - c). Only the vertical *kumiko* have alternate cuts (D22, B - B').

I could have done it the usual way in this project, but I decided to employ the straight lap joint. In a larger project, you'd have to be very careful not to cut the lap joints too tightly. If you do, the screen might bow. Mark the lap cut depth and saw the center lap cut. Put a small scrap of *kumiko* material in the notch to secure the piece, as you did on the vertical *kumiko*. Add a pencil mark to show which mark will become a blind mortise (D19, page 57).

Finish cutting the lap joints then cut the tenon. Don't forget the pencil mark on one side of the tenon cheek. The two outside *kumiko* have all the marks—those pieces will be the top and bottom of the leaf. On these, the two *kumiko* marks between the lap cuts will be the blind mortise (D22, b). Use the same mortise marking gauge to mark one side (photo 137). The depth of this mortise is about two-thirds the *kumiko* thickness. Chisel out the waste without damaging the mortise mouth. For this type of cut, Japanese typically use the *sokozarai-nomi*, but it is too big for this job. Instead, I transformed a small dental tool into a cleaning chisel. It works very well. After you finish this part and chamfer the tenons, you can assemble the *kumiko*.

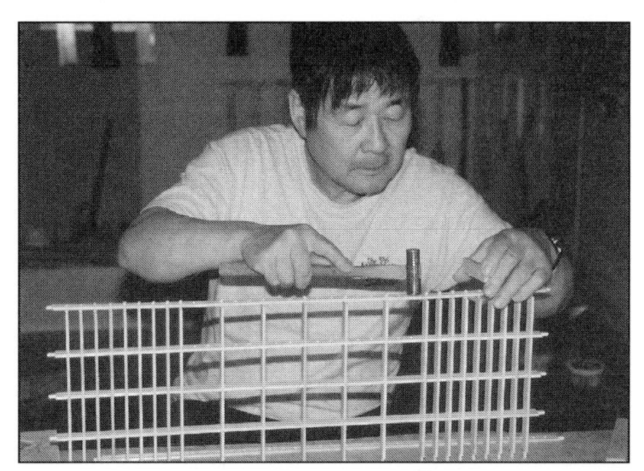

138 - 144 Assembling *kumiko*.

ASSEMBLING *KUMIKO*

The Japanese use a little rice glue, but flour glue will also do the job, as it will for any *shoji* project—the glue plays only a small role in the assembly. At this point in the project, I would not use glue for the *kumiko* frames. Only when I assemble the outer frame will I add glue, so that the pressure on the *kumiko* frame becomes even. After assembling the *kumiko*, put in the hemp leaf pattern (page 71).

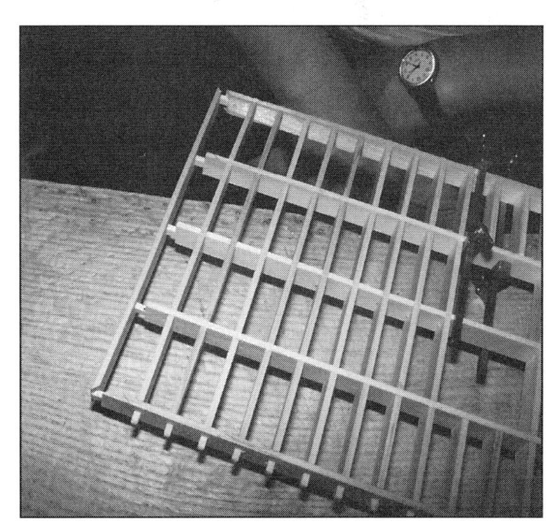

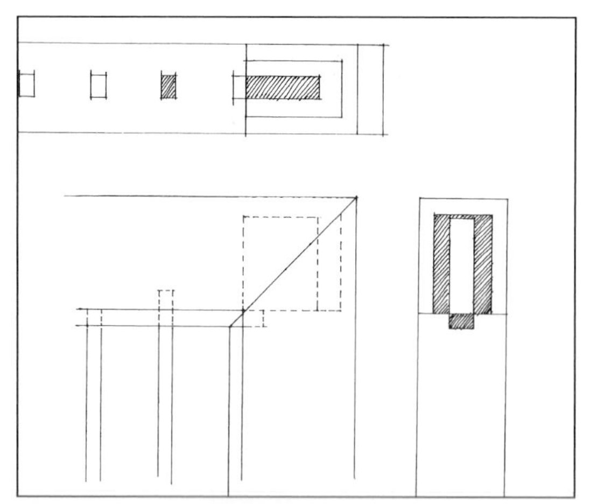

D28 Boxed mortise and tenon joint.

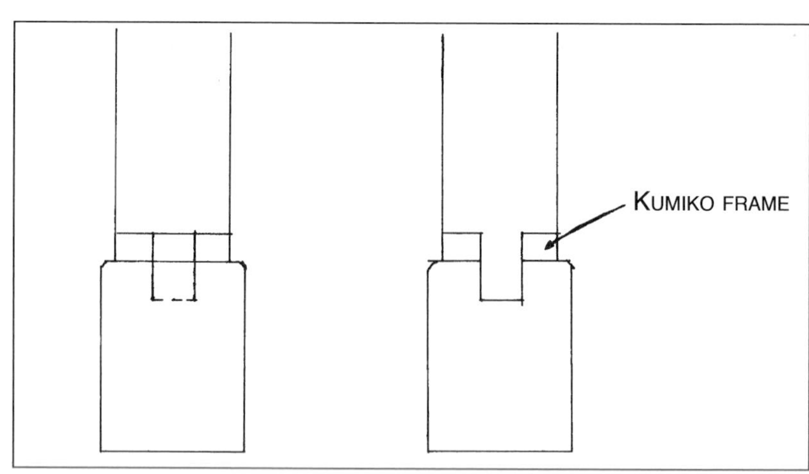

D29 Position the kumiko frame in the center of the outer frame.

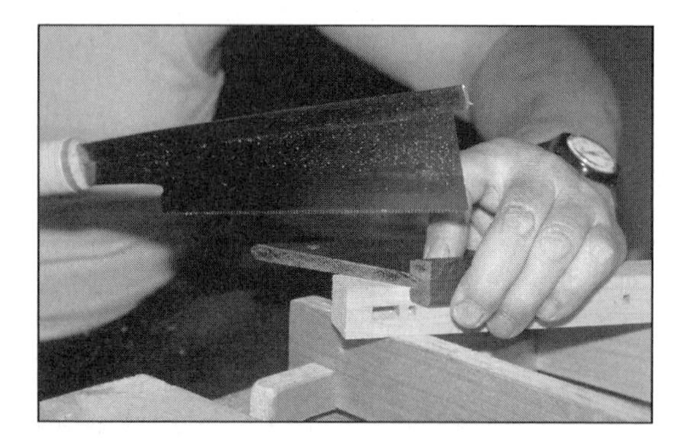

145 - 147 Cutting miter of vertical frame.

OUTER FRAME WITH MITERED BOXED MORTISE-AND-TENON JOINT

The next step is the outer frame. On this transom one doesn't apply paper or glass, since both faces are equally visible. Thus, you position the *kumiko* screen in the center of the frame (D29). Where you previously marked *kumiko* positions, mark all mortises and mitered ends with a mortise and tenon in the center of the frame. Make sure which *kumiko* marks will be mortised (D19, page 57). The depth of the mortise should be about 5/16" (8mm)—not less. Cut it with a chisel as deeply as possible, without damaging the mouth. Then use the cleaning tool to take out the waste. Continue in this way until the mortise is about 5/16" (8mm) deep. I use a depth indicator, which I made especially for this project. *Kumiko* mortises are very small and quite deep, so it is difficult to clean the bottom. *Shoji* makers in Japan use a square or rectangular metal rod that is slightly smaller than the mouth. The depth mark is indicated on the side of the rod. To be certain the depth is clean and correct, you simply pound the rod into the mortise, same as when making common *shoji*.

148 Upper part of boxed tenon has just been cut.

149 Ripping cheek of boxed tenon.

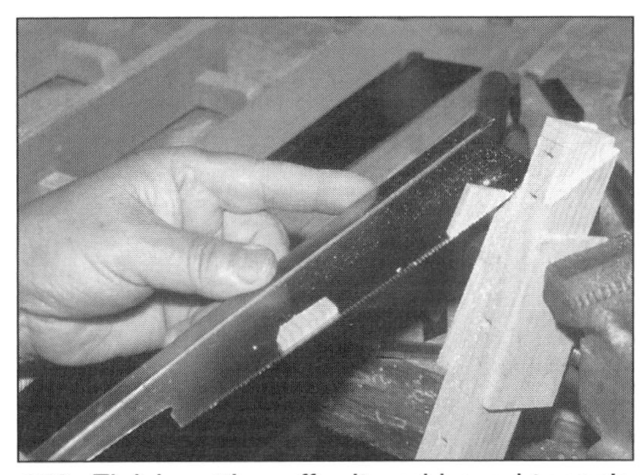

150 Finish cutting off mitered boxed tenon's cheek.

MITERED CORNER

Next will be the mitered corner. There are several types of mitered corner joints. The one to use depends on what project one makes and where it is used. I use the traditional Japanese corner joint, called "mitered and boxed mortise and tenon joint" (D20-21, page 58). It is quite intricate, but it is a very strong joint that will not loosen for many long years. Some ask, "It seems a shame to use such a beautiful joint where no one can see it. Why do you use it?" People often forget about the reason and meaning of joints. The essence of a joint is its strength and the trust you can place in it, not qualities that are perceivable by the eye.

In Japan this type of frame is a standard thickness. Therefore, this corner mortise and tenon are usually 3/8" (9mm) thick. However in this project, the joint needs more material for the side of the mortise and tenon, so they will be 1/4" (6mm) thick instead of 3/8" (9mm) (D28).

Each part of this joint contains both mortise and tenon—very different from the conventional mortise and tenon joint.

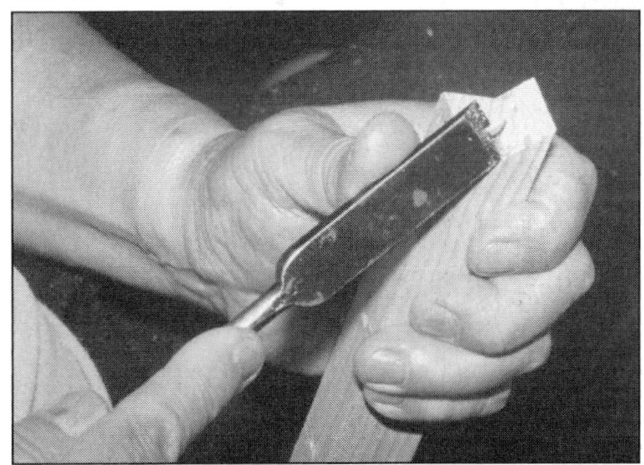

151 Chamfering the boxed tenon of vertical frame.

152 Finished vertical corner joint.

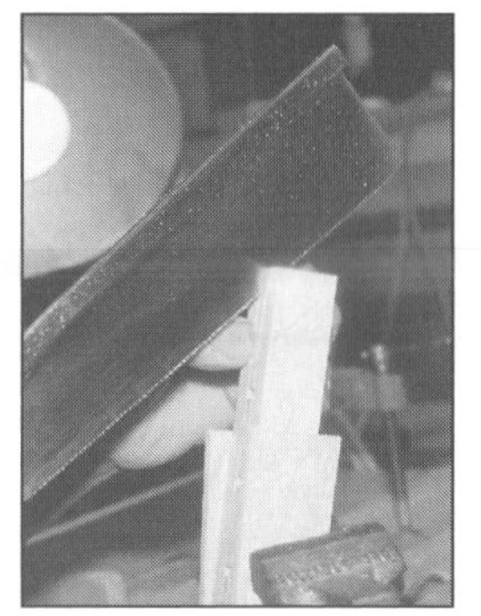

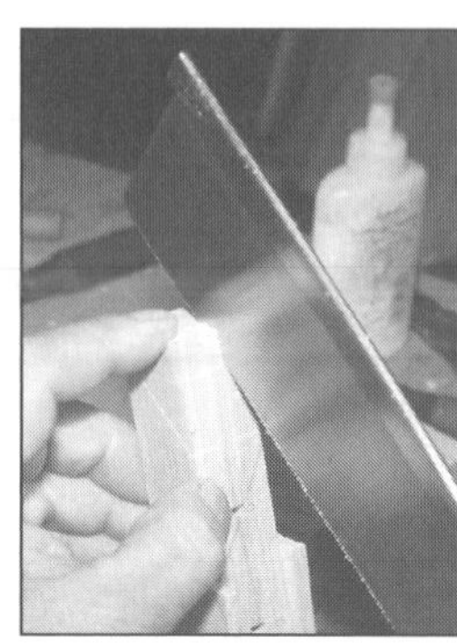

153 - 154 Cutting tenon/miter corner of horizontal frame.

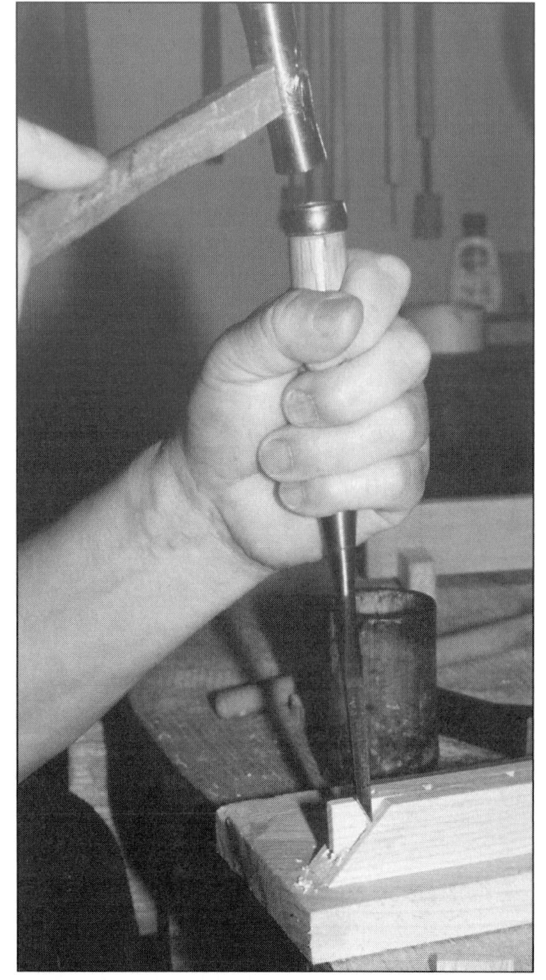

155 Mortising box mortise.

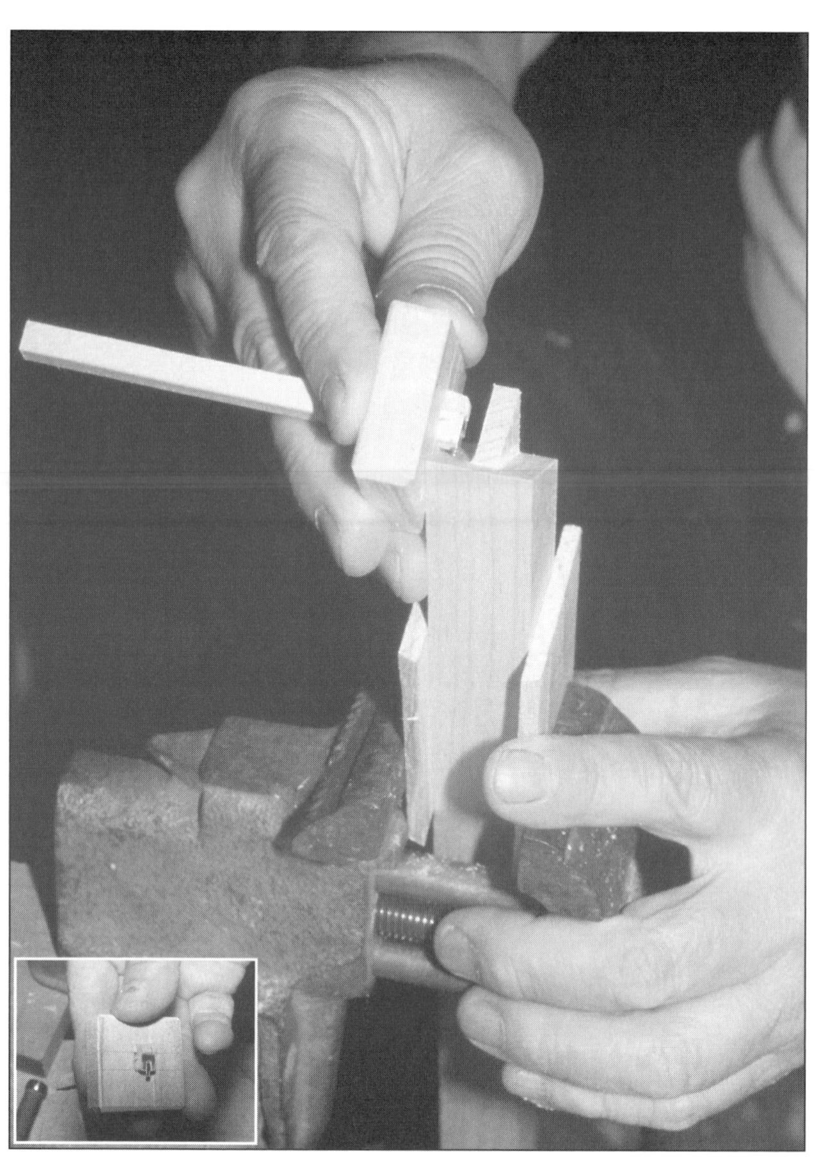

156 - 157 I made a special little end-pin marking gauge to mark the box mortise joint.

In making this joint, one section will be particularly hectic. You'll have to dig behind the 1/4" (6mm) tenon, which is a very small end-grain area. I first used a gimlet; if you decide to use a drill, be careful that the bit doesn't slip (photos 158-159). Then I managed to clean the space (photos 160-162). For this size joint, I make the tenon 1/16" (1.5mm) shorter than the depth of the mortise, which should be as deep as possible without going through the other side. If this is the first time you are making this type of joint, I'd suggest that you practice a few times. Prior to assembling it with glue, you should dry-fit the joint, even though Japanese usually don't dry fit, for they believe it takes too much time and if it is soft wood it will weaken the joint. After assembling this transom you can sand or plane the face.

158 - 159 Using gimlet to make space behind tenon.

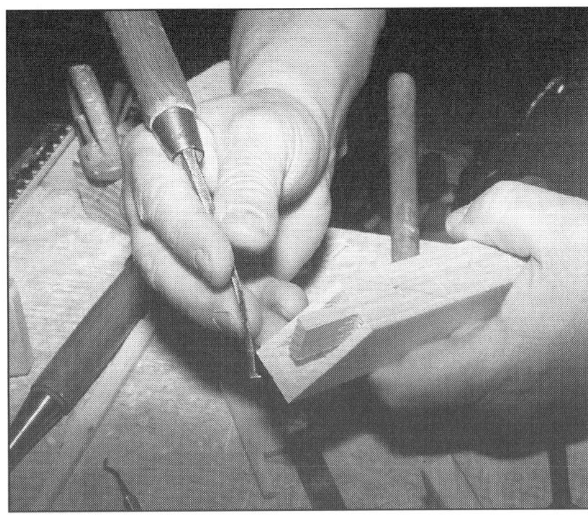

162 Cleaning mortise of boxed tenon with bottom cleaning chisel.

163 Adjusting size of frame's tenon width.

160 - 161 Cleaning mortise for boxed tenon.

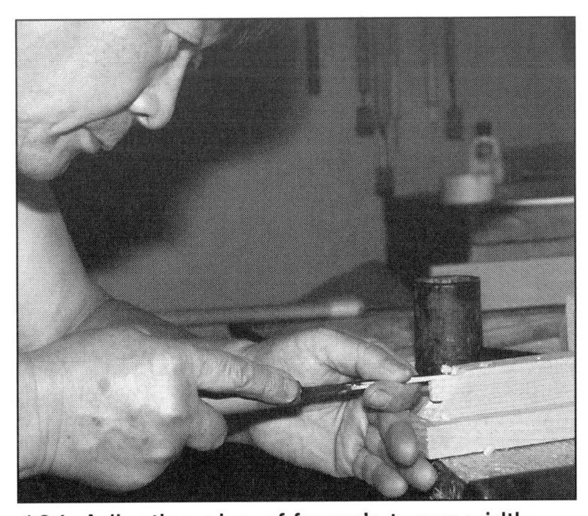

164 Adjusting size of frame's tenon width.

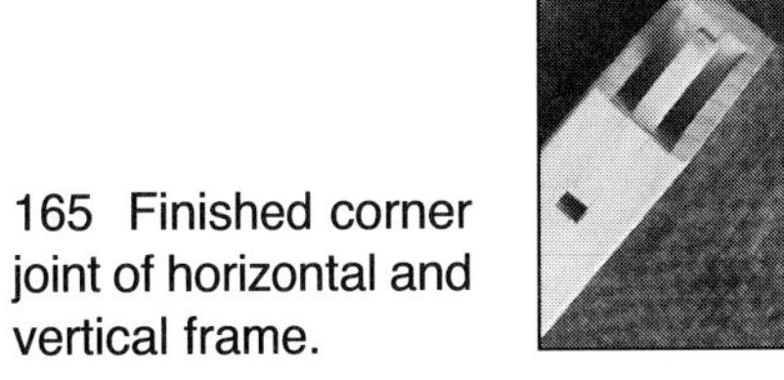

165 Finished corner joint of horizontal and vertical frame.

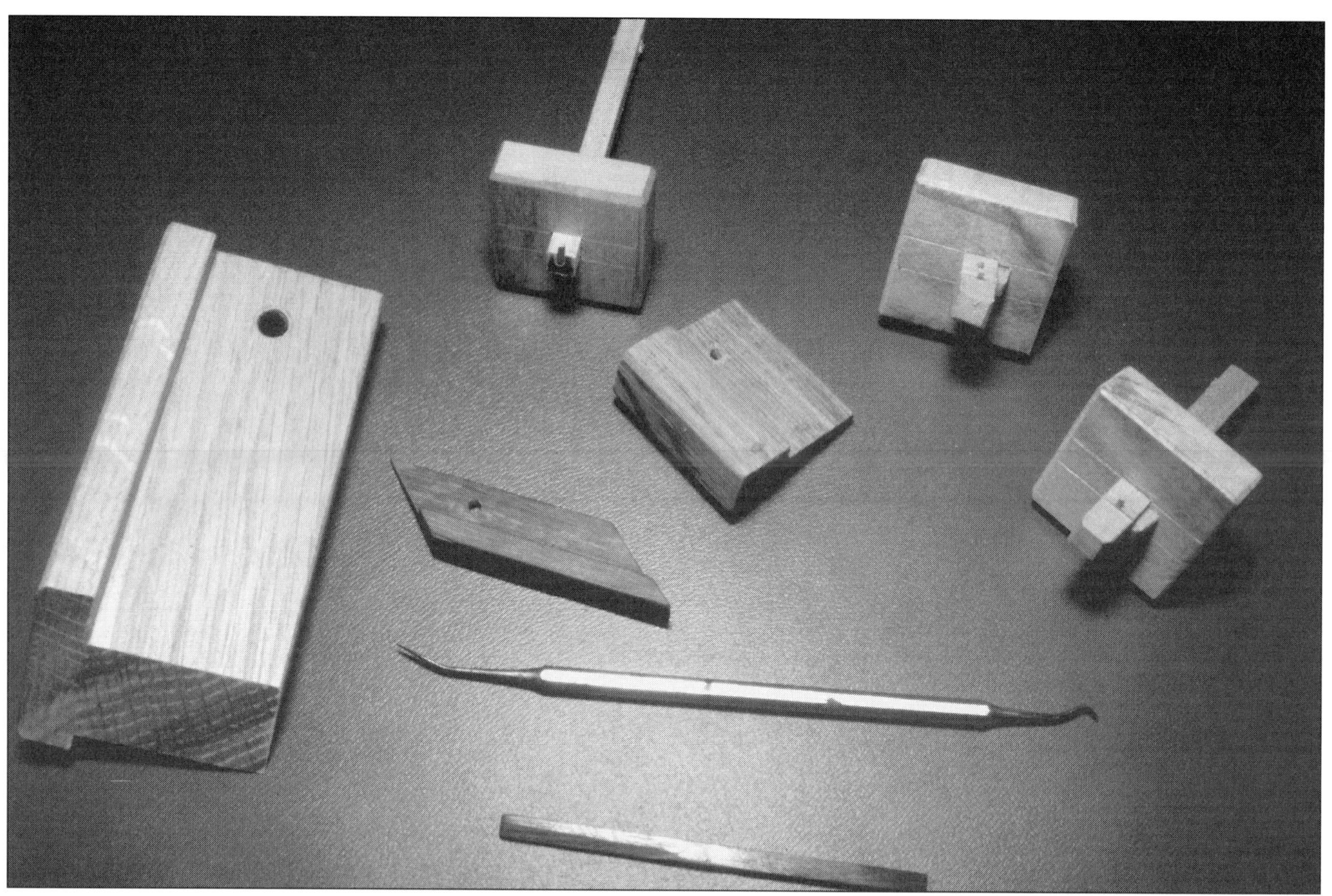

166 I made special tools for this project:
One small common marking gauge
One small end pin marking gauge
One small mortise marking gauge
One small 45 degree angle ruler
One small 90 degree angle ruler
One small marking knife
One small bottom cleaning chisel
One small pounding depth indicator rod
One small *uchinuki* (waste-cleaning stick)

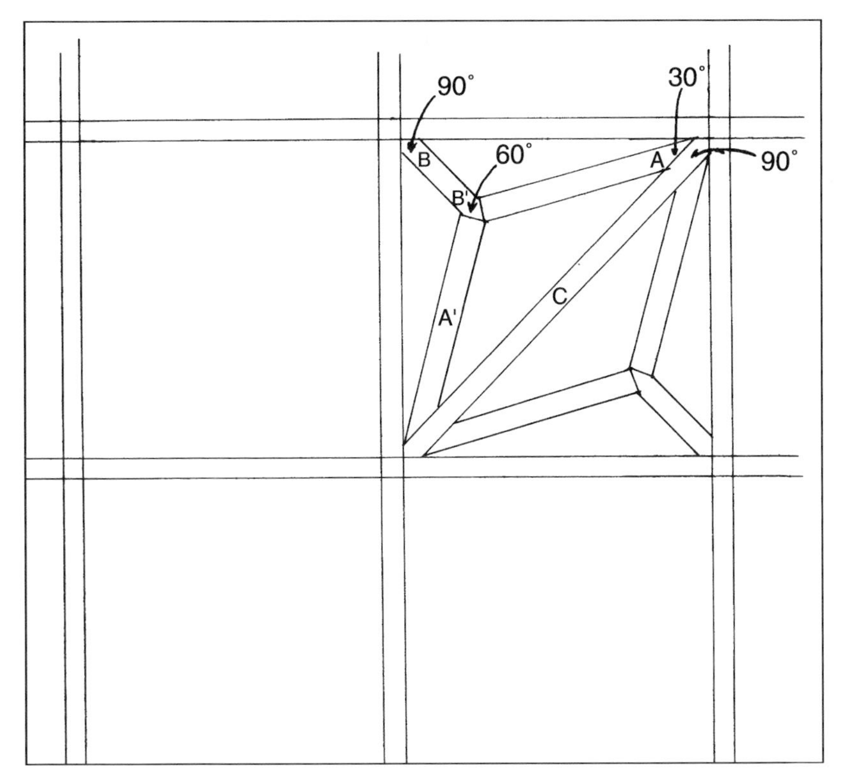

D30 The first step in making the decorative pattern is an accurate full-size layout to find the angles of the *kumiko*. Here the C piece is cut with 45-degree angles to fit into both corners. The A piece is cut with a 30-degree angle on both ends. It is hinged in the middle by being cut almost all the way through. The B piece serves as a key and has a 45-degree angle on one end and a 30-degree angle on the other.

167 *Asanoha*, or hemp leaf, pattern

MAKING *ASANOHA* OR HEMP LEAF PATTERN

Shoji are commonly composed of a simple series of squares or rectangles covered with rice paper, but the motif has also spawned several decorative forms of flower, leaf, or geometric patterns called *hanagata-kumiko*. Some of these patterns are quite complex and call for the very precise joinery of *kumiko* work. One of the simpler patterns is called *asanoha*, or hemp leaf. The pattern, although used most often for *shoji* work, can also be used for lampshades, under glass on a tabletop, or for a variety of other surface decorations. To form this pattern inside a square framework you need only three different wood parts, which come from the *kumiko* stock. One of those parts is carefully cut so that a kerf in the middle doesn't quite sever the workpiece—leaving a hinged piece that forms the sides of two triangles. Together with a pair of locking keys, the hinged pieces create a self-locking *kumiko* that's very pleasing to the eye.

You can build this pattern with the aid of two simple jigs. One guides a plane to cut accurate angles on the ends of the workpieces. The other jig is a small miter box set up for a 90-degree partial cut of the hinged piece (D31, page 72).

When assembled, the pieces fit together like a puzzle, with small keys providing pressure that holds the frame solidly in place. The method still takes some patience, even with the jigs, but with a little perseverance you'll achieve good results. This method is based on insetting the pattern into a square frame. For a

168 Cut a 90-degree angle, dead center on the hinged piece, with the aid of a miter box. Set a stop on the saw to achieve the precise depth so you won't cut all the way through. Two to three paper thicknesses of wood will be left to create a nice hinge.

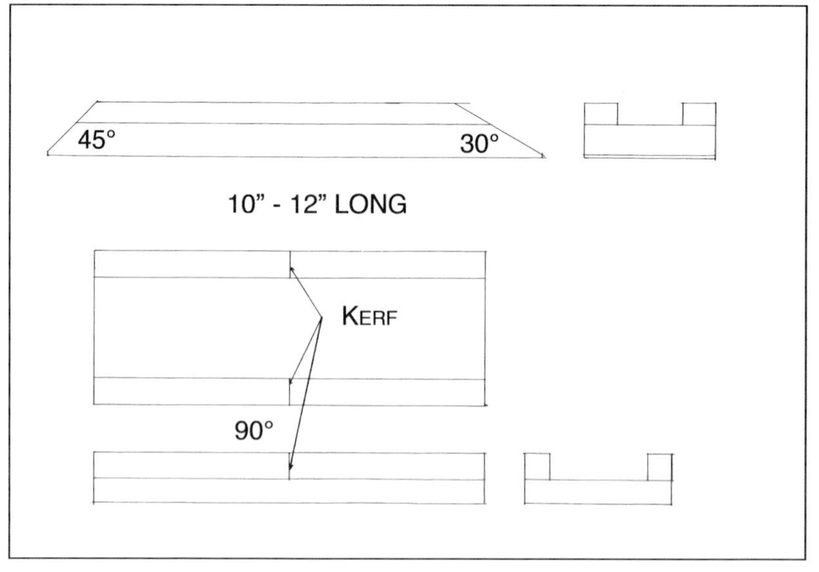

D31 These two jigs ensure the precise angles needed for the *kumiko* to fit properly and form the hemp leaf pattern. The top jig is used with a hand plane to cut the 30- and 45-degree angles on the ends of the three different pieces that make up the pattern. The bottom jig is used for carefully cutting the pieces that use a sliver of wood as a hinge.

169 Use both angles of the jig to plane the ends of the keys. One side will have two 45-degree angles, the other side will have two 30 degree angles.

170 The hinged piece with a pair of 30-degree angles.

first try, make up a practice square that's about 5" by 5" (127mm x 127mm). The *kumiko* should be about 5/16" (8mm) by 3/4" (19mm). The procedure naturally depends on the ability to achieve the precise angles on the ends of the *kumiko* and the frame pieces. The jig makes this possible, but the job still takes a sharp plane and some practice. The cuts are made toward you, a style of work that fits a Japanese-style plane better than a Western block plane. Either type of plane will work, however.

Before starting the work, assemble the jigs and cut some test pieces. It's especially important to make sure you can cut precise angles. Remember that the actual workpieces should start out a touch longer than the dimensions given in the layout. This will provide some margin for error and allow you to sneak up on the precise fit.

GETTING TO WORK

Make the jigs from any appropriate hardwood. Lay out both angled sides carefully so they are exactly 30 and 45 degrees. You can use the full-sized drawing and a sliding bevel to transfer the angles onto the sides of the jig. I cut the material carefully with a handsaw, staying a bit off the layout line. Use a sharp hand plane for the final cut. You can also make the jig with machine tools.

For the 90-degree kerf in the middle of the small miter box, use the same saw you plan to use for the actual work. I recom-

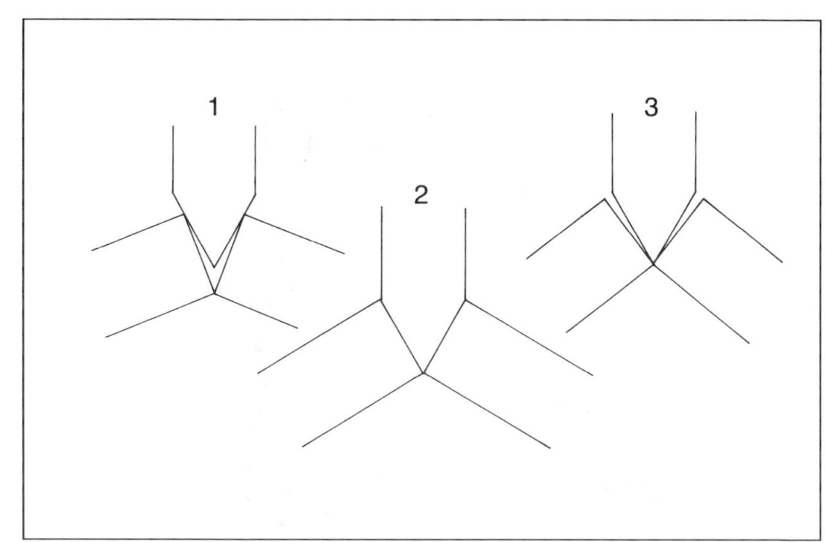

D32 If the three parts are all uniformly *kumiko*-thickness, then the imaginary triangle at 2 has to be an equilateral triangle with all three corners at 60 degrees. If any of these corners are not 60 degrees, then the key will not fit correctly, as at 1 and 3.

171 Lap joining the *kumiko* together.

172 The *kumiko* square is now ready to be made into one *asanoha* pattern.

173 Diagonal piece with a pair of 45-degree angles.

174 Fit the diagonal snug but not too tight, because the pressure might distort the frame.

mend using the thinnest saw you have. To keep the jig from sliding around in use, hammer wire brads into the bottom corners and clip off the heads, leaving a set of protruding points. Working from the layout drawing (D30, page 71), it's easy to see how the jig angles are derived

Angle of *Kumiko*	Jig
A = 30 degrees	30 degrees
Two sides cut, jig angle is half:	
B = 90 degrees	45 degrees
B′ = 60 degrees	30 degrees
C = 90 degrees	45 degrees

Once the jigs are made, you're ready to go to work. Prepare the *kumiko* on a tablesaw with your best sawblade; finish up with a hand plane or thickness planer. Make sure to allow enough stock for a few offcuts. Start with the diagonal piece, C, which fits into the 90-degree corners. It is mitered at the tip with a pair of 45-degree angles. Work carefully with the plane and jig so you can arrive at an exact fit. The piece should fit snugly, but not so tightly that the pressure distorts the frame angle. Next, cut the A-A′ hinged piece. This piece needs a very precise length, so I cut the material a little bit longer. Use your jig to plane the 30-degree angle on one side of all 24 pieces (photo 170). In this case, I always make a couple of extra pieces. I fix a stopper in the jig for precise length and finish the other 30-degree angles.

For the next step you'll need the miter box set for 90 degrees (photo 168). You'll cut the hinged piece in the middle and use a stopper to make sure you cut all of them dead center. This cut is

175 When the hinged piece is inserted into the pattern, the two sides of the kerf will open to form a 60-degree angle.

176 Place the key into its pattern and it will lock the hinged piece into place.

177 One *asanoha* is now completed, for the keys are snugly placed and hold the pattern firmly together.

178 All pieces needed to make an *asanoha* pattern.

crucial to the procedure, since the pieces aren't cut all the way through. By using a stop on the saw, it's possible to set up the miter box to achieve the precise depth. Set the stop by laying two or three sheets of notepaper inside the miter box. Then, on the paper, insert the *dozuki* saw and clamp a strip of wood to both sides of the sawblade (photo 168, page 72). When you remove the paper, the setup will cut the kumiko to a depth that's about two or three sheets short of the full thickness. This will provide just the right amount of wood to bend easily and form a hinge. When inserted into the pattern, the two sides of the kerf will open to form a 60-degree angle (photo 175). If you wet the other side of the kerf, it will bend more easily and not break off.

Piece B fits into that angle. It has two 30-degree angles that meet to form a 60-degree angle. The other end of B, however, is cut to the same angle as C because it fits into the corner of the square frame. This piece will serve as the key that provides tension and locks the design together. First cut it at a 60-degree angle (30 degrees x 2). Leave it a little longer, then work it down by planing away tiny shavings until it makes a snug fit (photo 169, page 72). When the last key is cut, it should fit snugly and hold the framework tightly together.

After dry-assembling the frame and getting the pieces to fit satisfactorily, place a dab of glue on the ends of the pieces and do the final assembly. The Japanese use rice glue, which doesn't so much fasten the pieces as help them to stay in place. White glue, yellow glue, or a thin hide glue work better for hardwood.

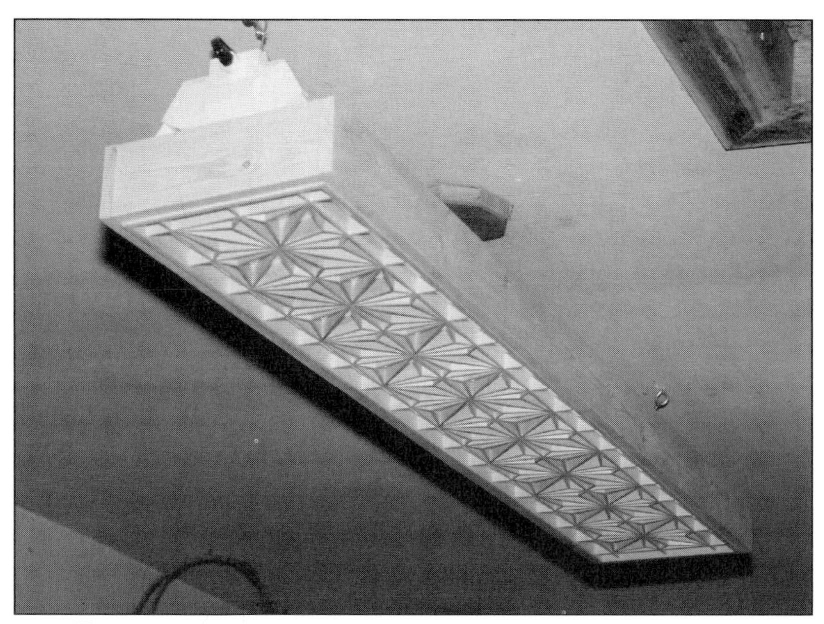

179 The *asanoha* pattern can be found not only on a transom or *shoji* but also on different types of lamp shades. Here, the pattern and *Shoji* paper soften a bright fluorescent light.

180 - 181 Night lights with *asanoha* pattern.

The Japanese Mortise-and-Tenon

When woodworkers from any part of the world build cabinets and furniture, frame houses, or make doors, chances are they will use a mortise-and-tenon joint somewhere in the project. This fundamental joint consists of a tenon with a shoulder on each side that fits into a mortise of the same size.

Most people regard a mortise-and-tenon joint as permanent or semi-permanent, so they don't think of using it without glue, wedges, or pins. But there are applications in cabinetry and carpentry where glue, wedges, and pins should not be used—for example, where you want a strong joint that is reversible. Doors, for example, sometimes need to be taken apart for repairs. Some framing joints are held together from the sheer weight of the building, so they also do not need wedges or pins.

The Japanese way of making a mortise-and-tenon joint is different from the Western approach, which has been well described in Western publications. In Japan, the *tategu-shi* (sliding-door maker) and *daiku* (carpenter) use different methods to make a mortise-and-tenon joint. The carpenter employs the *taiko,* or drum-style through-mortise, for framing work. The door-maker uses the *tsuzumi,* or shoulder drum-style mortise. The key to the *tsuzumi* joint is in the way the tenon is squeezed between the walls of the mortise and locked in place. This is why softwoods are used for *tsuzumi* joints; hardwood will not compress and spring back to shape as easily (D33).

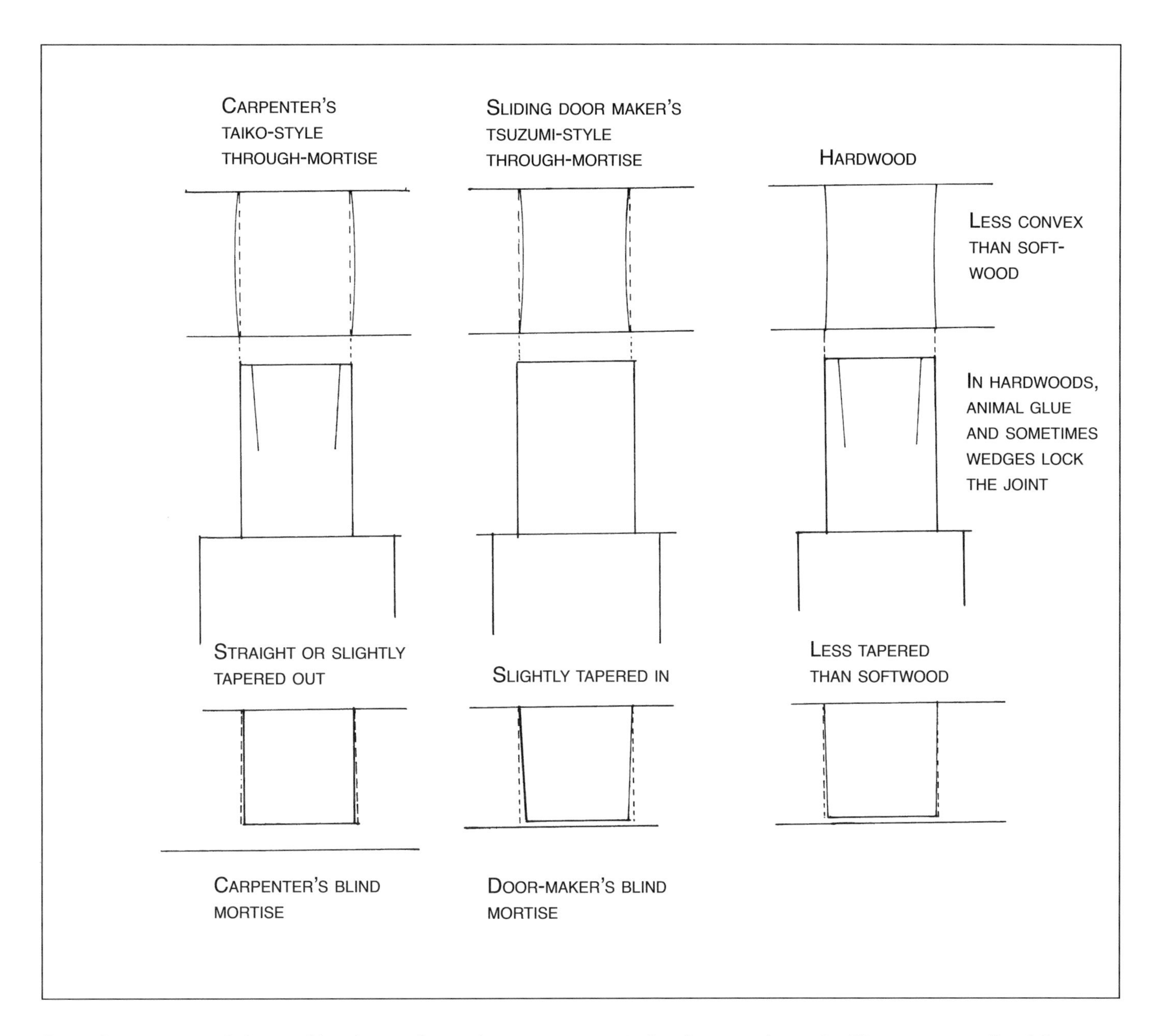

D33 Carpentry and doormaking have always been two very separate trades in Japan. Different joints and tasks are used for building and furnishing houses.

***Taiko*: carpenter's through-mortise**

Carpenters use this through-mortise for house frames. The concave walls of the mortise allow easy assembly for large timbers. Tenons are sometimes cut for wedges, or pinned through adjoining members.

The end grain of mortise walls is cut out more in the center to prevent friction with the tenon as it is driven into the mortise. The tenon width is cut exactly the same size as the mortise length.

***Tsuzumi*: Doormaker's through-mortise.**

Doormakers use this through-mortise for many types of exterior doors and panels. The convex walls of the mortise compress tightly around a straight tenon as the joint is driven together. Water in the rice glue helps the tenon swell back to shape, locking it in place.

Mortise walls at the center are slightly narrower than the length of the mouth. The tenon is slightly wider than the length of the mouth.

Blind mortises

Carpenter's blind mortises are straight or slightly tapered out. These joints are used on heavy timbers. They are held together by their own weight, or sometimes pinned.

Doormaker's blind mortises, used in room-divider screens, are slightly tapered in. Straight tenons are squeezed to fit the shape.

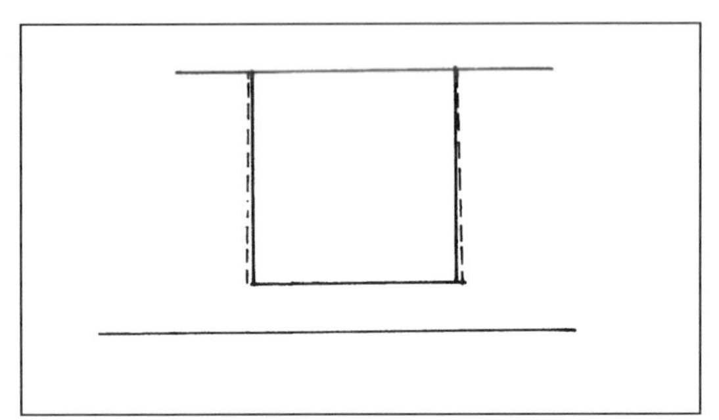

D34 Carpenter's through and blind mortise.

CARPENTER'S JOINTS ASSEMBLE EASILY

In laying out a through-mortise, the carpenter makes both mouths of the mortise the same width as the tenon. But the end-grain walls of the mortise are slightly concave in the middle—enough to avoid friction as the tenon is driven home and snugged into place. Usually the carpenter cuts the tenon 3/8″ to 1/2″ (9mm to 12mm) longer than the depth of the mortise (the excess is trimmed flush after assembly if it interferes with another part of the structure). Sometimes the carpenter cuts the tenon for wedges, especially when framing the foundation of a house. A carpenter may also pin the tenon in place and chamfer the end of the tenon on all four edges (as often seen on Western Arts-and-Crafts furniture). The carpenter always rips the tenon cheeks first, then cuts the shoulders, because any overcut at the shoulder would affect the strength of the tenon. The second cut meets the initial kerf, which forms a natural stop to prevent the saw from cutting into the tenon.

When the carpenter cuts a blind mortise, the mortise and tenon are the same width and thickness. The end grain walls are cut straight and square to the edge of the workpiece, or the mortise is cut slightly more to relieve any friction when joining the large pieces of wood. The length of the tenon is 3/8″ to 1/2″ (9mm to 12mm) shorter than the depth of the mortise, depending on the size of the joint. The carpenter uses an ink line to mark the center and a bamboo pen to mark the mortise and tenon, because the ink lines are easier to read and bamboo is more durable than pencils.

Carpenters score the cheek cuts of the mortise along the grain with a butt chisel, and dig out mortises with a thick chisel that is narrower than the mortise. A thin, wide, and long paring chisel is used to clean up the cheeks.

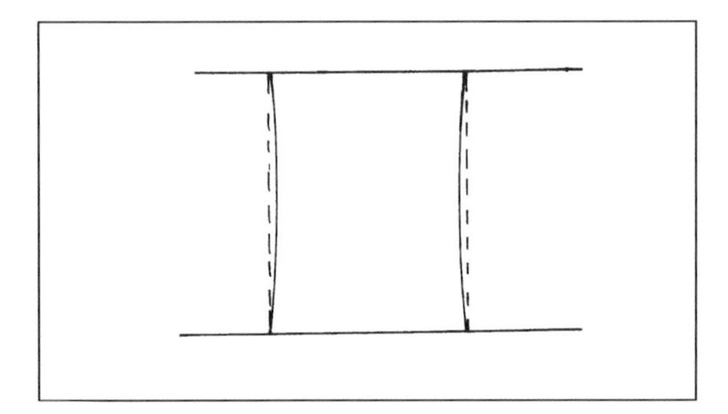

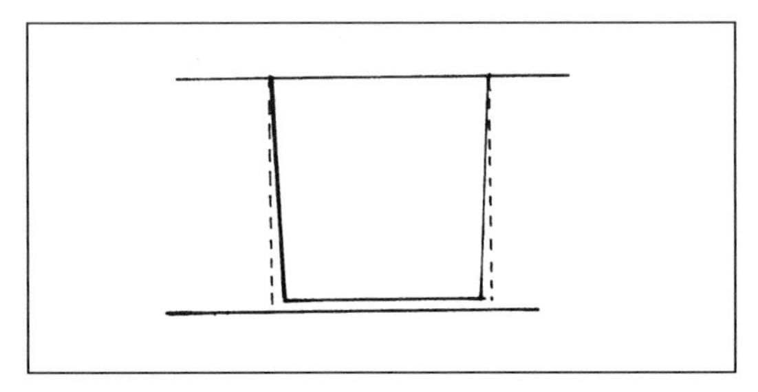

D35 Sliding door maker's through and blind mortise.

TOLERANCES ARE FINER FOR SLIDING DOORS

The door maker's through-mortise in the *tsuzumi* style is slightly convex at the middle of the end grain walls (D33, page 77). These joints are traditionally used for exterior doors. Carpenters always make the tracks for sliding doors and screens for the doormaker, and the track size is mostly the same throughout Japan. The stiles are always about 1 1/4″ to 1 3/8″ (32mm to 35mm); the door maker uses 1/4″ and 3/8″ (6mm to 9mm) chisels for most of the work. The door maker uses raw materials called *itawari*: a 1 7/16″ (36mm) thick by 6′ (1830mm) long plank of wood that is ripped and dressed to make the frame for the sliding doors and screens.

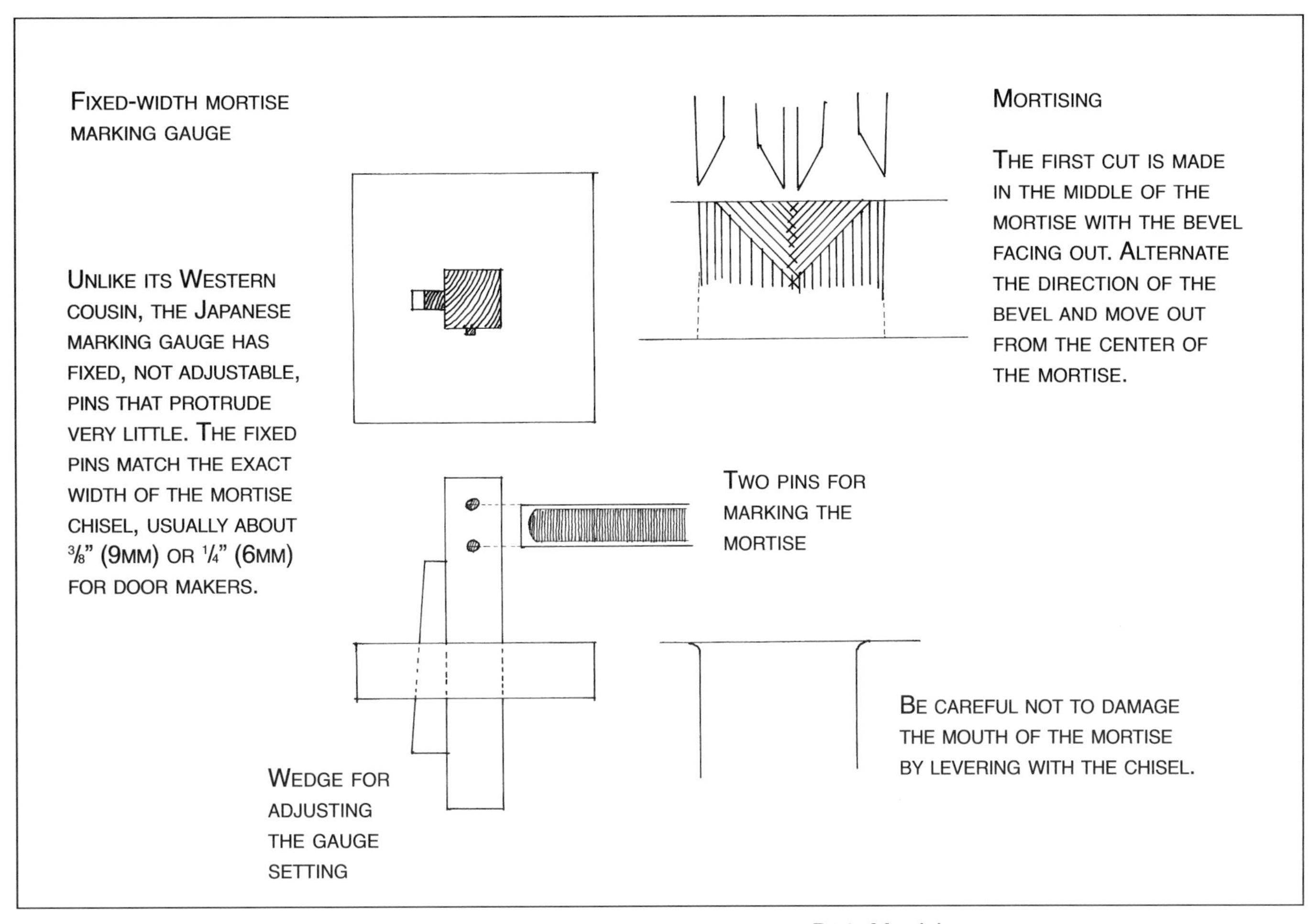

D36 Mortising.

The Japanese door maker uses a marking gauge with fine pins to mark the mortise width line. Unlike Western adjustable marking gauges, Japanese gauges are set for one chisel size—either 1/4" or 3/8" (6mm or 9mm). A marking knife scores the cut on either end for length. The door-maker chops out the mortise by striking with an iron hammer, not the wooden mallet used in the West. One begins in the middle and works out toward the ends, turning the chisel around with every strike. That way, most of the waste is pushed out or loosened.

The final mortise cuts are made with the flat back of the chisel facing toward the end line. The door maker does not use a butt chisel to cut with the grain along the cheek line, because the width of the mortise chisel is exactly the same as the tenon. Using a butt chisel might change the width of the mortise. For a through-mortise, half the depth is cut from one side, then the workpiece is flipped and cut from the other side—the same method often used in the West. A piece of oiled cotton in a bamboo pot is used to lubricate the chisel as the door maker digs out the mortise. To push out the waste, the door maker uses a *uchinuki-nomi* (blunt strike-through chisel) (D37, page 81).

182 Dip the chisel into the oil pot to lubricate it. The cotton wadding is soaked with vegetable oil.

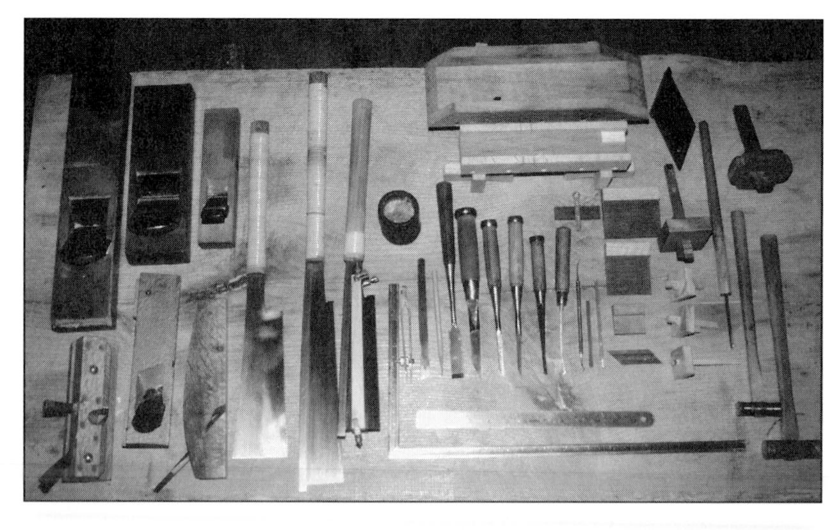

Door maker's tools
183 - 184 Toshio Odate and his traditional Japanese tools used for making a *shoji*.

The *tategu-shi's* tenon shoulders are always cut first, before the cheeks are ripped, because a slight overcut won't show once the joint is assembled. The tenon is wider than the mortise by 1/16" (.5mm) or less, depending on the type of wood being used. For a through mortise, the length of the tenon is 1/8" to 1/4" (3mm to 6mm) longer than the mortise depth, and the four corners of the end are chamfered. For more detail on tenons, see page 38.

Rice glue is usually used for assembly: It lubricates the pieces, making them easier to join, and serves as a filler around the rough fibers of the cheek. This rice glue, called *sokui*, is made by cooking rice then mashing it into a paste (page 44). After a tenon is squeezed past the narrow halfway point of the mortise, moisture from the rice glue helps the tenon expand back to shape, locking it in place. After assembly, the extra tenon is sawed off and then planed flush, removing any marking lines. Later on, if a door needs repair and it's necessary to take the joint apart, all that is needed is a slight hammer blow to break the glue bond.

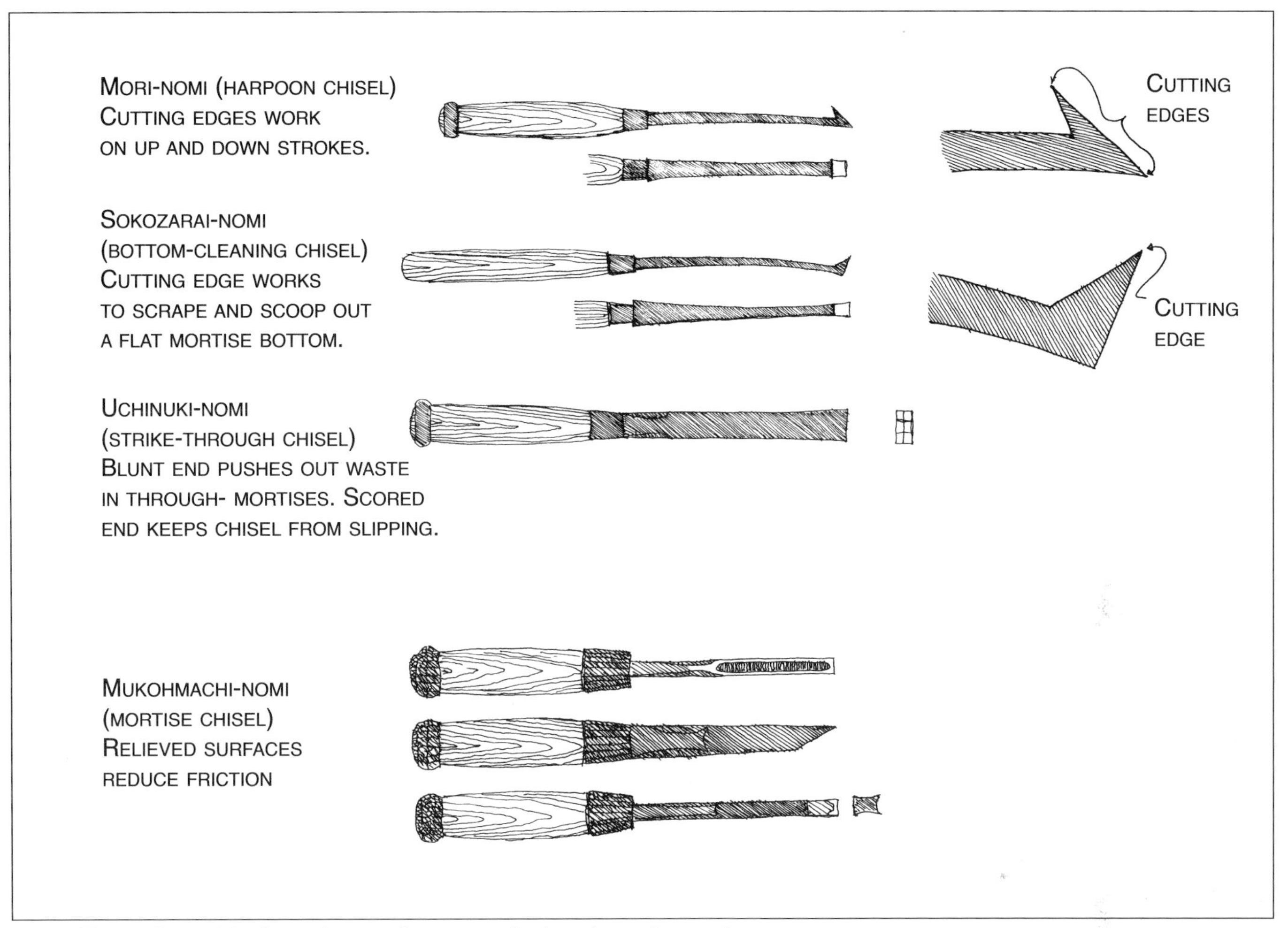

D37 These four chisels make up the arsenal of tools a doormaker uses to cut blind or through-mortises.

SEEING LIGHT THROUGH THE WOOD

In Japanese woodworking, end grain and wedges are considered unappealing and inelegant, especially when used in furniture and doors. Blind mortise-and-tenon joints are much more appropriate for interior work. But they are not so easy to use because of the scale of the work. Most Japanese rooms are very small compared to American rooms, so screens and door frames are smaller.

A Japanese room is encircled by *shoji* (translucent-paper sliding panels) or by *fusuma* (opaque paper sliding panels). Wide-faced stiles for interior panels do not work, because their proportions are not soothing to the eye of people sitting on a *tatami* (grass mat) floor. The face of the stile should be 7/8" to 1 1/4" (22mm to 32mm) wide; for strength, the tenon must go almost all the way through the mortise (photo 186, page 82). For this type of work the Japanese have special chisels: the *mori-nomi*, or harpoon chisel, and the

185 Blind mortises must be very deep without breaking through, and slightly tapered for a tight fit. A depth gauge made from a small *kumiko* scrap indicates when a blind mortise is deep enough.

186 One can see light coming through the blind mortise.

187 The *mori-nomi*, or harpoon chisel, cuts on the up and down stroke, and lifts the chips out of the mortise.

188 The *sokozarai-nomi*, or bottom-cleaning chisel, scrapes the bottom of the mortise flat.

sokozarai-nomi, or bottom-cleaning chisel. After a few strikes of chopping the mortise with a mortise chisel, put the flat back of the harpoon chisel flush against the mortise cheeks, then tap down the tool and quickly jerk it up. Its hook catches the chips and either loosens them or clears them out. Chipping alternately with the mortise chisel and the *mori-nomi*, proceed quickly until final depth is approached. Then you slow down and gauge the depth with a piece of *kumiko* material cut to the depth and including a shoulder (photo 185, page 81). Very gently score the remaining wood with a chisel, and use the *sokozarai-nomi*, a thin, goose-necked tool with a small spade bent at its end, to level the bottom of the mortise. I have never seen any Western equivalent to this tool. It is not tapped with a hammer but used with one or two hands like a scraper. Scrape with the *sokozarai-nomi* until the shoulder of the depth gauge touches the corner of the mouth. When the mortise is finished, light can be seen through the bottom.

The end-grain surfaces of a blind mortise should taper slightly inward just like the socket of a chisel handle or a tapered sliding dovetail joint, so it presses evenly on the tenon from the mouth to the bottom of the mortise. The tenon's length is about 1/16" (1.5mm) less than the depth of the mortise. If the wall were straight, you would only have pressure around the mouth—not a strong joint. The tapered shape also makes this joint easier to take apart, for making repairs later.

Another strictly enforced rule in the fundamental education of a doormaker's apprenticeship is not to damage the top edges of the mortise by using a chisel as a lever to remove the waste. I do not hear about this much in Western woodworking, but I was hit and yelled at many times during my apprenticeship for this transgression. We never use the corner of a mortise for leverage, or even accidentally press down on it with the bevel side of the chisel. One reason is obvious: The damage will leave a space that can be seen after the joint has been assembled. Equally important, because the tenon is short, such damage would weaken the joint.

PHOTO COURTESY MIYA-SHOJI, NEW YORK

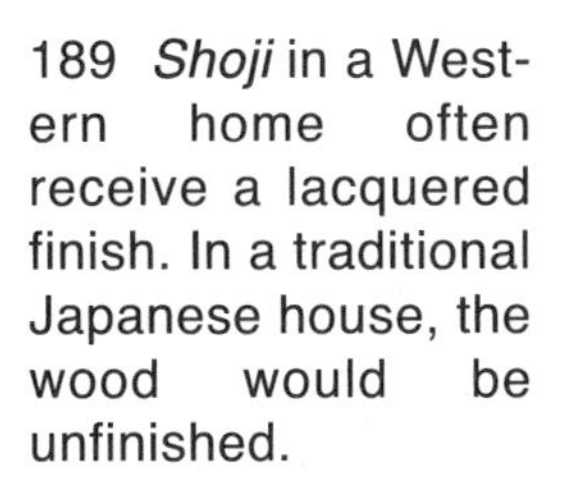

189 *Shoji* in a Western home often receive a lacquered finish. In a traditional Japanese house, the wood would be unfinished.

PHOTO COURTESY MIYA-SHOJI, NEW YORK

190 Sliding screens here are used as window coverings as well as doors.

The Story of a Regrettable Scar

The *tategu-shi*'s blind mortise goes very deep, leaving only about 1/16" (1.5mm) thick material at the bottom. To clean and flatten the bottom is often quite a tense task. I made several mistakes in my life, and on one of those occassions I really felt a cold sweat. I will share this story with you.

I had just begun to work at customer's homes, without my master. Then, I would only do very basic and simple jobs like the common *shoji* or rain doors.

One day, I left for work alone. This customer, who was a farmer, had beautifully colored Japanese cedar which had been sliced and dried since his grandfather's lifetime. Most of the boards had already been used for a project and only a small amount was left. "What do you think, is there enough for four *shojis*," he said. "I will see," I replied.

There were a few reasons why he wanted to use this beautiful wood for a common *shoji*. One was that the *shoji* could be seen from the street. I checked the wood and thought there would be just enough material for the four *shojis*. The material preparation went very well and I was well ahead of schedule. Everything was marked. I then placed four stiles together, and started mortising. As I was cutting one of the mortises for the middle rails, the chisel had almost reached the bottom. But my depth gauge indicated to scrape another 1/16" (1.5mm), so I did with the *sokozarai-nomi*. Beautiful red-colored shavings came out from the mortise. I then inserted the depth gauge again, showing no change.

So I scraped again, and more colored shavings came out. I reinserted the depth gauge, but still no change. I was puzzled, but I scraped one more time. Surprised, I noticed white shavings mixed with a bit of red coming out. Suddenly a flash of lightning hit me. I turned over the stile, and indeed there was a regrettable scar. I had been careless. I looked into the mortise, and noticed a small step on the end grain. Somehow I had created it, and the depth gauge had rested there the last couple of times I had used it. I had no extra wood to replace the stile, I almost panicked. Just at that moment the customer came from the fields. With a big smile he said, "Sorry Tossan"—this was what people called me then—"my wife has to stop by the grocery store, so lunch will be a little late." "Already lunch time...no problem," I replied, and smiled back. I tried my best to keep the problem to myself.

After lunch everyone went to the fields. The beautiful stile had an ugly hole, 1/8" (3mm) wide and 3/8" (9mm) long. I made very strong rice glue and planed a couple times on the ugly hole, which then became a little bigger. I made a very thin veneer-like piece, matched the grain, then glued it on the hole. Before the end of the day I had planed it very carefully. Nobody could see the hole but I knew it was there. I finished the four *shojis*. All came out quite well. It wasn't my record speed but the fourth day was just a half a day. I installed them and cleaned the place before noon. I just hoped that the little veneer would stay on there forever.

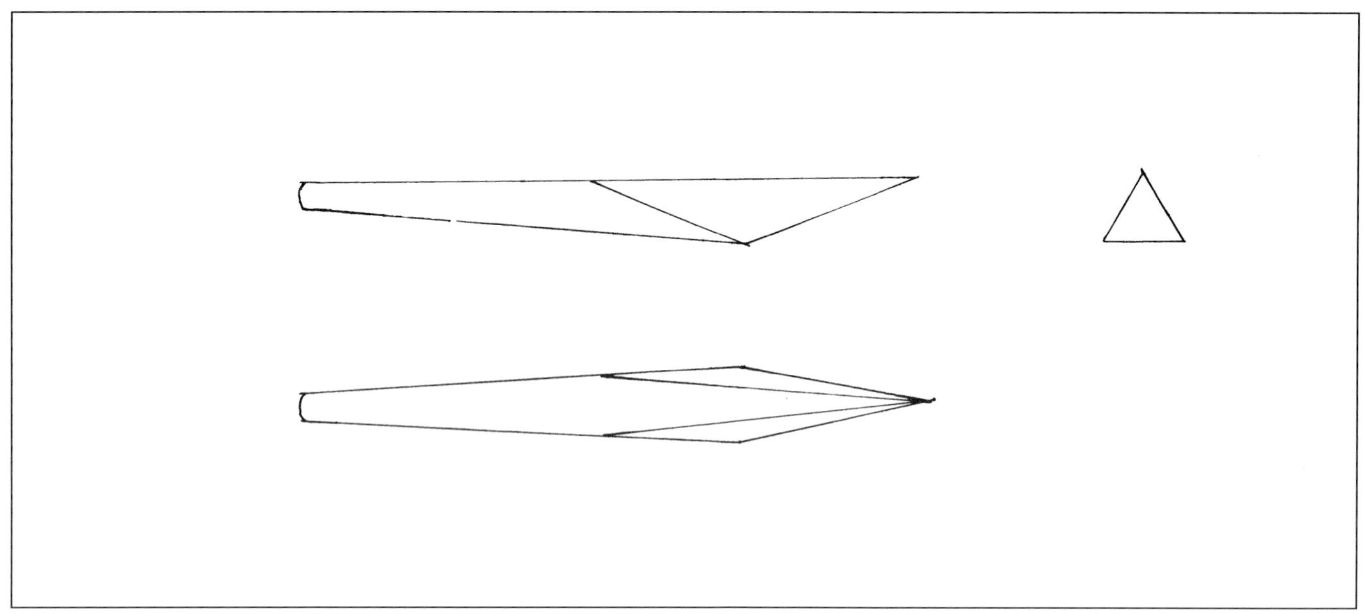

D38 *Nori-bo*, glue-mashing stick.

Japanese Rice Glue

Japanese woodworkers and Westerners have quite different attitudes toward glue. Both appreciate the idea that the strongest glue is best, but one must also consider the purpose of using glue. When the Japanese woodworker makes a joint such as a mortise and tenon, the joint itself—not the glue—is what holds the two pieces of wood together. The doormaker will make the strongest joint by using his best skills and knowledge. The glue then simply assists the joint. Japanese commonly use rice glue, which is easily made, dries transparent, and does not discolor for a long time. In addition, its consistency and drying time ease application and cleanup.

Glue is also used when Japanese woodworkers place two pieces of thin, soft boards together, which is called *hagi,* or edge-to-edge gluing. In this case the glue holds the boards together. This edge-to-edge gluing technique is most visible in projects like the hipboard of a *shoji* or the center of a panel door. Customers as well as craftsmen are well aware that the joint should not come apart, but from time to time this does happen. When it does, it is a disaster for the craftsman, and the craftsman makes his best effort to correct the disaster. Sometimes customers supply a board with a very interesting figure, which is beautiful but very difficult to join because the grain runs off the edge so the wood has a mixture of side grain and end grain. End grain doesn't glue well, and is liable to move and twist in all directions. In this case, the craftsman does his best and can only cross his fingers and hope that the joint will have a long life. If for some reason the craftsman revisits the customer's house, the first thing he will do is peek out of the corner of his eye at the *hagi*, for he wants to see if the seam has been motionless. Naturally, the craftsman doesn't want the customer to perceive his worry. I was a door maker in a small village, and when the *hagi* came apart the village people would remember the incident for quite a long time. Thus, keeping edge-to-edge joints together became a craftsman's pride.

Hence, first the craftsman must understand the board's character. Then, with knowledge and skill he must plane the edge to the best configuration for the particular wood. Finally, he uses the strongest rice glue. The best quality of glue comes from well-cooked rice, not too watery nor too hard, but just right! I don't know for certain, but when I was an apprentice, I heard that the best rice glue is also the most deliciously cooked rice. A door-maker's apprentice thus learned at an early age how to cook rice for both the best glue and a delicious meal.

Here is how I learned to cook rice for glue. Start with regular rice (not converted). I always use a short-grained variety of rice. Wash it with cold water, rinse until the water is clear, and ignore the so-called enriched part. Drain the rice overnight in a colander (in Japan, these are made of bamboo), and cover with a cloth to keep the moisture in. In the morning take one part

191 To make rice glue, start with regular short-grained rice.

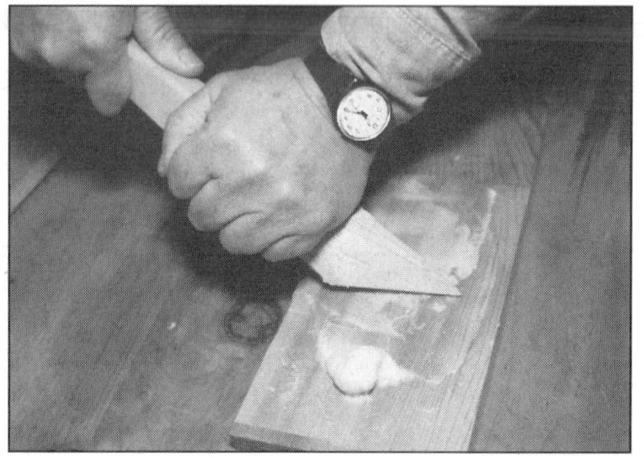

192 - 193 Preparing rice glue, *sokui*. Glue is used for keeping joints together, and as a lubricant during assembly.

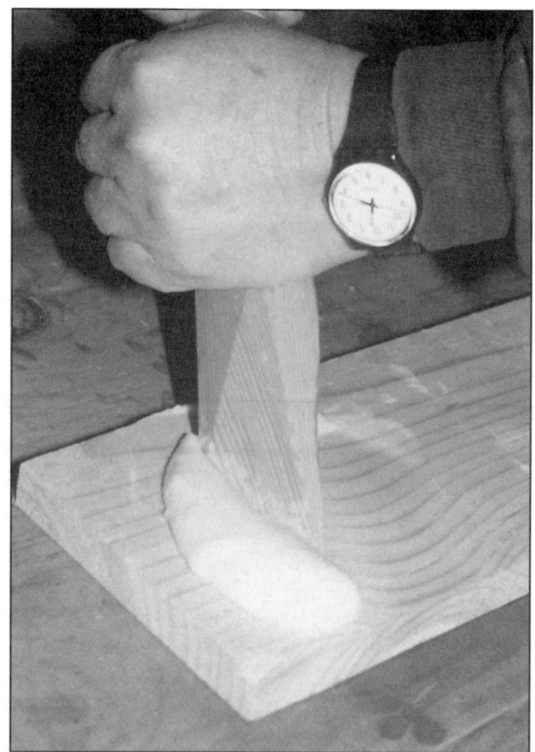

195 This glue is ready to use.

194 Mash the glue to an even consistency, then dilute it with a little water.

rice and one part water and put them into a thick, cast-iron pot with a heavy cover. Bring the water to a rapid boil. When it nearly boils over, turn the heat down to the lowest setting and allow the rice to simmer for about 15 to 20 minutes. Then shut off the heat and let it sit for another 10 minutes. Try to get the rice kernels well-cooked, tight and not mushy.

In the shop, prepare a glue-mashing surface by selecting a smooth board or a small sheet of plywood. Clean one side. Then make a mashing stick, called *noribo*, of soft wood (D38, page 86). Take the rice to the shop in a small covered bowl and put a small amount of rice on the work surface. Bring the flat part of the mashing stick down on it and work the stick back and forth with a rocking motion. Slightly lift the leading edge of the stick on each stroke, and continue mashing until the rice is smooth and pasty. Remove any bits of debris with the stick to the side and proceed mashing until you have enough glue. Usually this consistency will be a bit too thick for most uses, so you must add a little water to thin it down, but do not make it too thin. If your rice is soft and mushy it will be easier to mash and thin out, but it will not have much gluing strength. However, if you proceed as outlined here, the glue will be thin and smooth and will have great strength. This rice

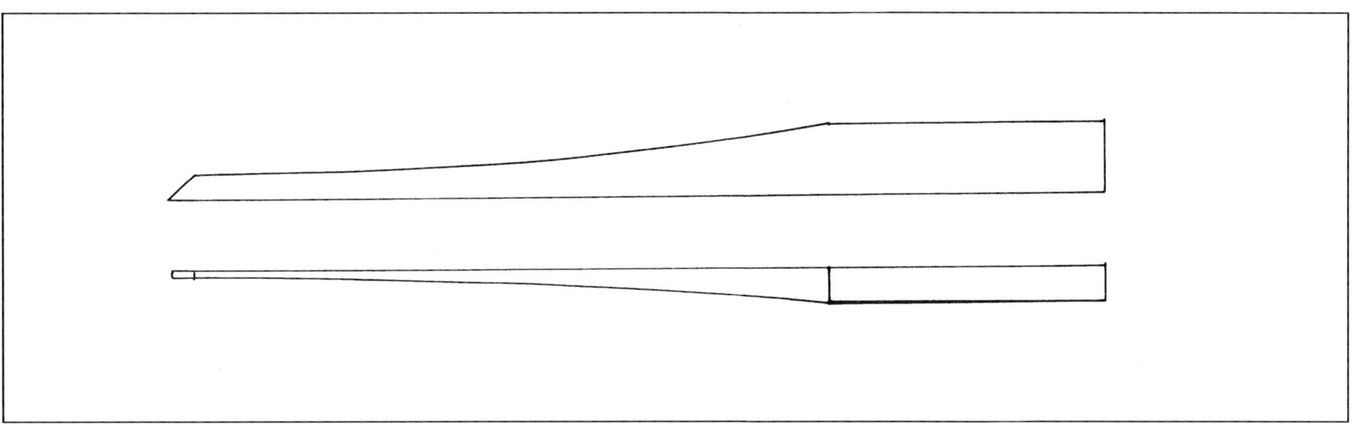

D39 Glue spreading stick.

glue is called *sokui*, but today not too many people know about *sokui* or *nori-bo* (the mashing stick), for the process is now so mechanized in Japan.

To apply rice glue to mortises or half-notched *kumiko* joints, the door maker uses a glue-applying stick made from leftover *kumiko* material. To apply the glue to the edges of boards, the door maker uses the glue-mashing stick. When applying glue to mortises, apply a little glue only to the walls, not the bottom. If you put too much glue on or if it gets on the bottom of the mortise, the glue will act as hydraulics do, and the tenon shoulders will not fit tightly in the mortise. When applying glue to the edges of the panels, try to create a little heap in the center of the edge. When applying glue from both sides of the edge, it will be easier to form this heap. The reason for this is simple. When you put the edges together, you will not create air bubbles. Of course you should also think of this when you use yellow glue.

Excess glue can be wiped off immediately with a damp rag, or you can wait 4 to 5 hours for the glue to dry. By then it will be easy to scrape or plane off the squeezed-out glue. Scraping dried rice glue won't hurt a plane or chisel blade as it would if one had used yellow glue.

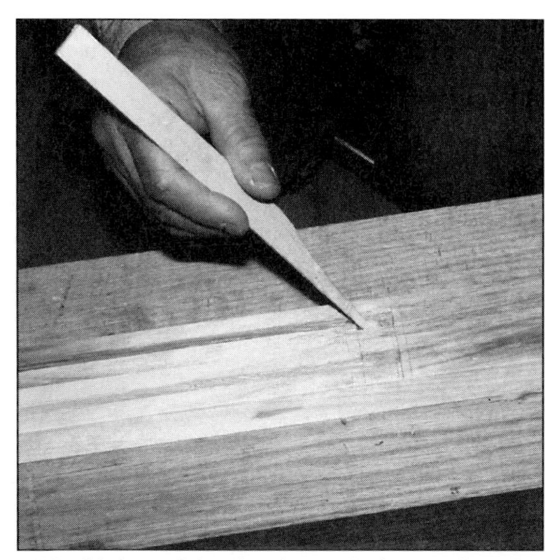

196 Gluing stick, simply made from a *kumiko* scrap, puts glue into a mortise.

197 Gluing edges of boards with the glue-mashing stick.

198 Here is a pair of common *shoji* seen from the paper side. In Japan the paper side would face out from the room. Note the handles let into the paper, explained on page 93.

Shoji Paper

The *shoji* paper draws in not only the brightness of the light, but also its warmth and softness, creating a very serene and peaceful atmosphere. In the 7th century, two men contributed to the making of paper in Japan. A Korean Buddhist priest introduced the paper-making method and Prince Shotoku suggested using the bark of the mulberry tree. The foundation of Japanese handmade paper production was then set. Since then, many different provinces have produced different qualities of paper. As early as the 12th century, *Mino* was established as the great paper-making province, and thus the paper produced there was called *mino* paper. Around the 17th century, the most popular paper for daily use was called *hanshi*, and it was produced in almost every paper-making province—where this name came from, I do not know. Westerners call Japanese paper "rice paper," even though Japanese papers have nothing to do with rice.

Most *shoji* paper comes in tight rolls. However, special handmade papers are likely to come in sheets measuring 23 1/4" x 36 3/4" (590mm x 934mm). Rolls of paper come in *mino* size—a 28 cm (11") wide roll—and *hanshi* size—a 25cm (9 3/4") wide roll. Some have watermarks and some are plain, and some also contain prominent coarse fibers that pattern the surface like

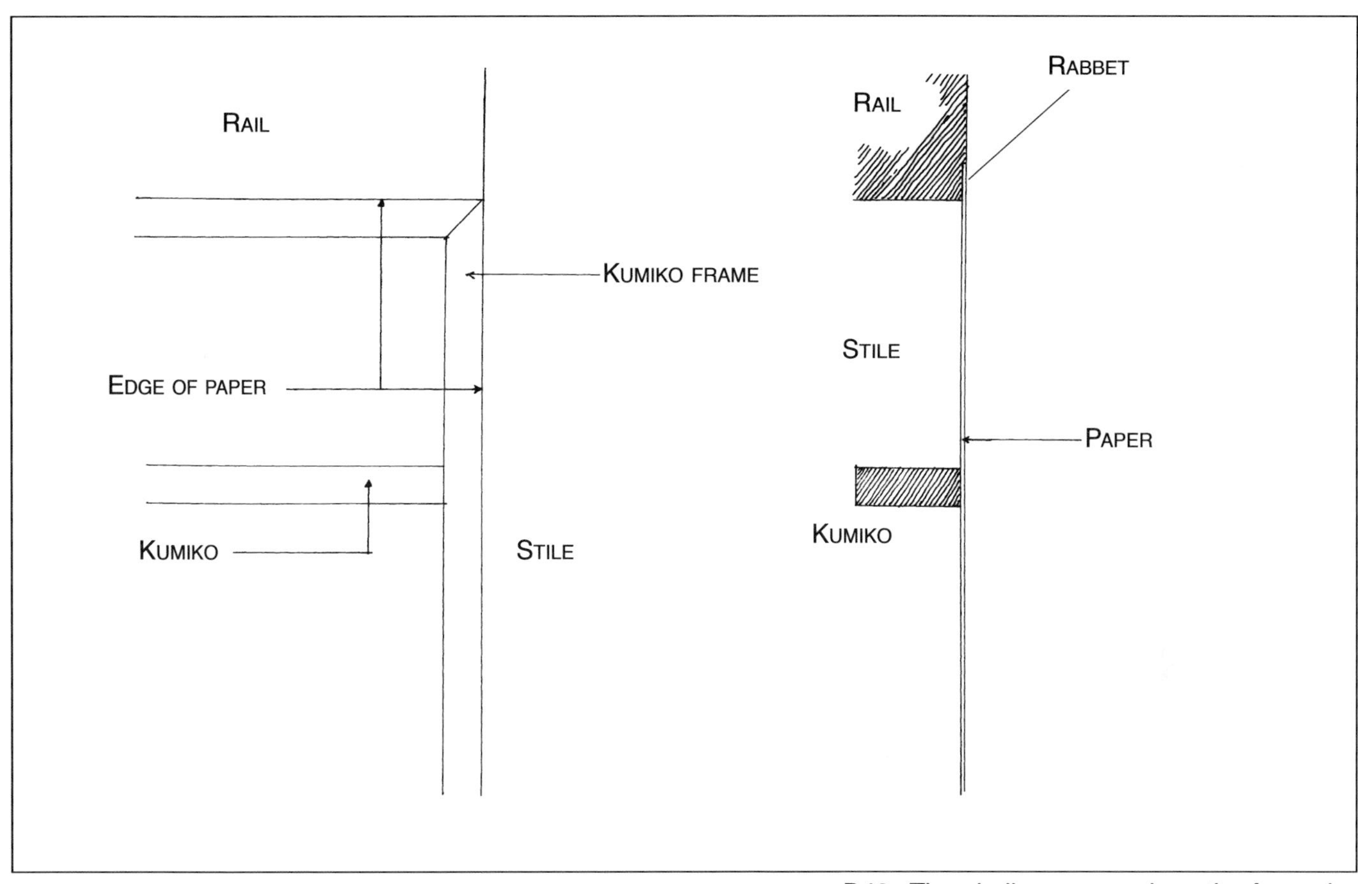

D40 The *shoji* paper overlaps the frame by the thickness of one *kumiko*.

abstract drawings. Watermarks are commonly seen in the form of a pine tree and needles, a bamboo and its leaves, a plum tree and its blossoms, and the chrysanthemum. All these are celebratory symbols in Japan. These watermarks suggest imaginary gardens close by the *shoji*, which cast their shadows on the paper.

Today these papers come in seamless rolls, but when I was young I vaguely remember the paper having an overlap seam of about 1/8" (3mm) every 2 1/2 feet to 3 feet (750mm to 900mm). I recall seeing this paper being applied to the *kumiko* with the seams carefully scattered around as if a beautiful pattern were being designed. The seams were never aligned one next to the other.

APPLYING THE PAPER

Rolls of *shoji* paper have a smooth side which is on the inside of the roll. It should face out when the paper is applied to the *kumiko*. The horizontal strips of papers are pasted to the *kumiko* with starch glue or wallpaper paste. As a rule, when overlapping the paper on the frames, cover the stile and middle rail just one *kumiko* thickness. Sometimes craftsmen make a little fancy detail at this point by adding a *kumiko* frame or they rabbet the *shoji* frame the width of one *kumiko* thickness and the

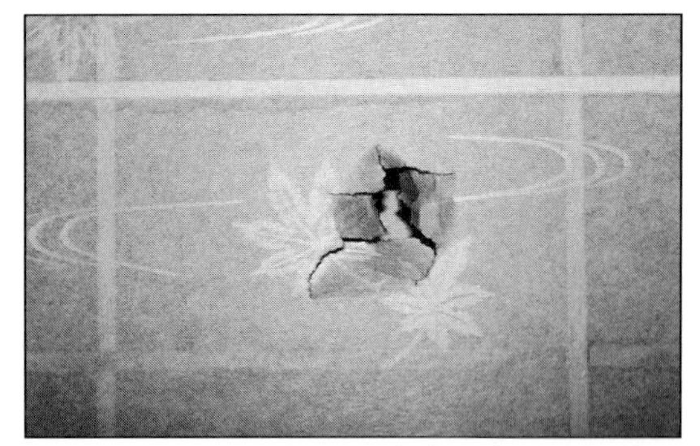

199 When paper is torn during the year, simply mend it by pulling the torn parts together.

202 A touch of nature is added to your home with the presence of a chrysanthemum and ginkgo leaf.

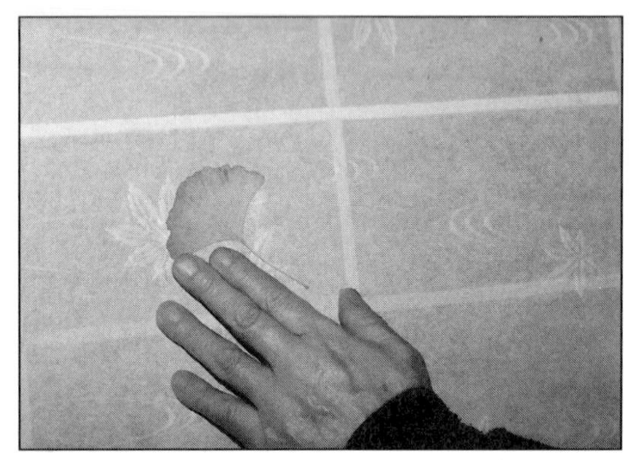

200 I am applying, on the pulled together *shoji* paper, a ginkgo leaf with starch glue. This will create the effect of a watermark or shadow on the paper.

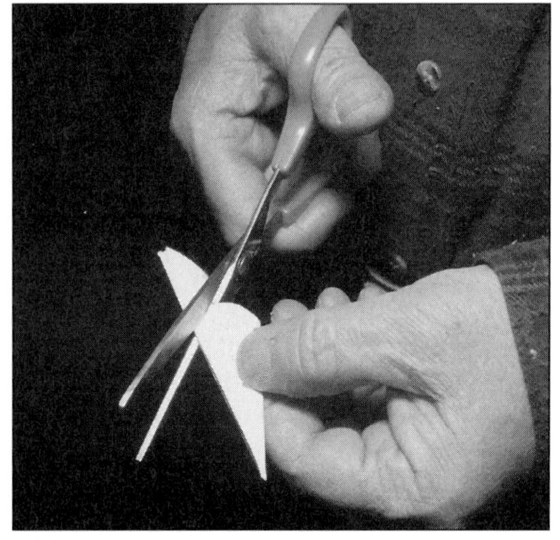

201 Now I will add the chrysanthemum by taking a small piece of *shoji* paper, folding it diagonally, then refolding the triangle three times in half from its center. I choose the size of the flower that will cover more than the torn area and cut the wide side into half a circle. Before you cut the original flower make a few try-outs!

depth of about one paper thickness (D40, page 91).

Use a wide, thin brush to apply the paste lightly, then tap it flat so the paste stays on the *kumiko* without creating a mess on the sides. Put on the strips of paper starting from the bottom of the *shoji* and overlap on the *kumiko* like shingles, so the seams will not collect dust. Paper is traditionally applied by the householders and customarily changed on the last day of the year. The year-old paper is changed to bright white paper for New Year's day, signifying a fresh new start. When you use paper with watermarks, be careful not to put it on upside down. Place the roll left to right, which is usually the right way, but you should still check it carefully.

MENDING OLD PAPER

Old paper is easily removed by moistening it. When paper is torn during the year, simply mend it by pulling the torn parts together. Then cut a new *shoji* paper in the shape of a chrysanthemum and cover the ripped area. Sometimes people insert a ginkgo leaf or a maple leaf to create a watermark or shadow effect.

203 I choose the height of the handle on the *kumiko* side of the frame and I'm now applying, with a light touch, flour glue onto the *kumiko* frame.

204 After applying the glue I gently place the rougher side of the paper onto the frame and stile.

205 Both papers for the future handles have been applied and are waiting for the starch to dry.

SHOJI HANDLES

The *shoji* has a front and back. The paper side is the front, which faces outward, and the *kumiko* side is the back. The front side is completely covered with paper—there is no opening for your fingers to push the screen open and closed. So you cut the paper out of one opening to make a handle. So the handle doesn't go through, paste a new piece of paper on the back side.

Next to the stile, choose the desired height of the handle. Cover one opening with *shoji* paper on the *kumiko* side (back), with the smooth side of the paper facing out. Then slice the front section of the paper diagonally to form an X. Paste the triangles onto the inside of the *kumiko* and on the backing paper (photos 208-210, page 94). This is a very common way to make *shoji* handles, however, often I see the handle opening filled with a 3/8" (9mm) wooden panel, decorated with a small motif in its center.

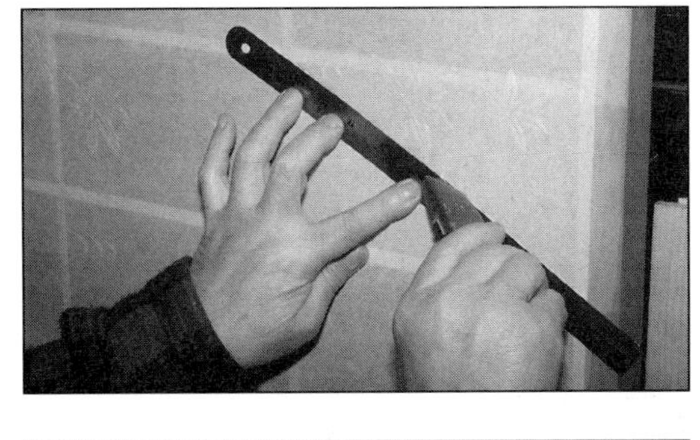

206 - 207 Slice the front section of the paper diagonally to form an X shape.

208 Fold and shape the sliced paper so it covers the *kumiko* and stile.

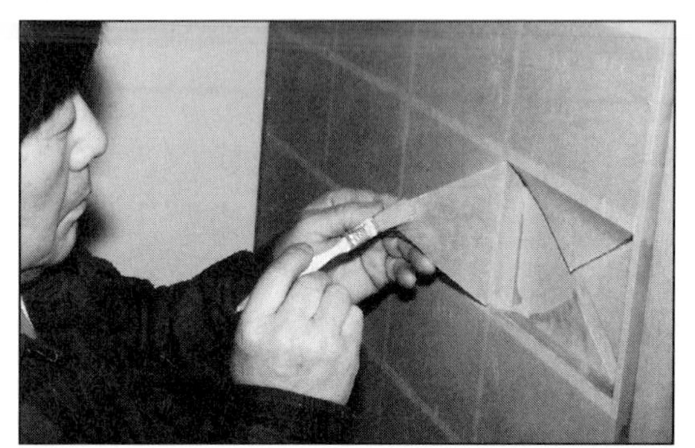

209 Paste the triangles onto the inside of the *kumiko* and on the backing paper. As a final touch spray the handles with water, so the paper will contract and dry evenly. Try not to wet other parts of the screen.

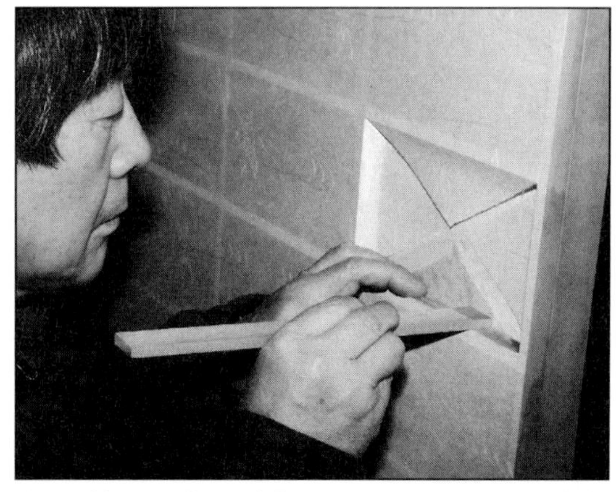

210 Use a flat stick to press the paper fold down squarely.

211 - 212 *Shoji* handles. You might be surprised that paper handles can be so sturdy!

CHANGING TIMES

Up to the 1950s, especially in the countryside, most *shoji* were made in *mino* size, which use three vertical *kumiko*, nine horizontal *kumiko*, a hipboard, and five strips of 11" high *mino* paper. Then, however, many people started using *hanshi*-type *shoji* to bring more light into the room. *Hanshi shoji* use three vertical *kumiko* and eleven horizontal *kumiko* with a shorter hipboard and six strips of paper. At the same time, *shoji* in contemporary Japanese houses used six strips of *mino* paper and completely eliminated the hipboard. One just adjusted the top and bottom rails to fit the six strips of paper. More changed in the 1960s. Besides the traditional *mino*- and *hanshi*-size rolls, paper companies made rolls wide enough to be pasted on vertically and in one piece. This opened up many possibilities in the spacing and patterning of the *kumiko*, which had always been carefully positioned to accommodate traditional-size paper. This kind of change creates freedom in *shoji* designs, but also raises the question of what will happen to the hundreds of years' worth of traditional, beautifully balanced, designs created by ingenious craftsmen under strict limitations. All of this is very much part of the Japanese woodworking culture.

Photo courtesy Miya-Shoji, New York

213 - 214 In a Western house, you might have to mount the *shoji* track on the wall, as shown in this example. These screens take advantage of modern paper, which allows the *kumiko* to be spaced any distance apart.

Photo courtesy Miya-Shoji, New York

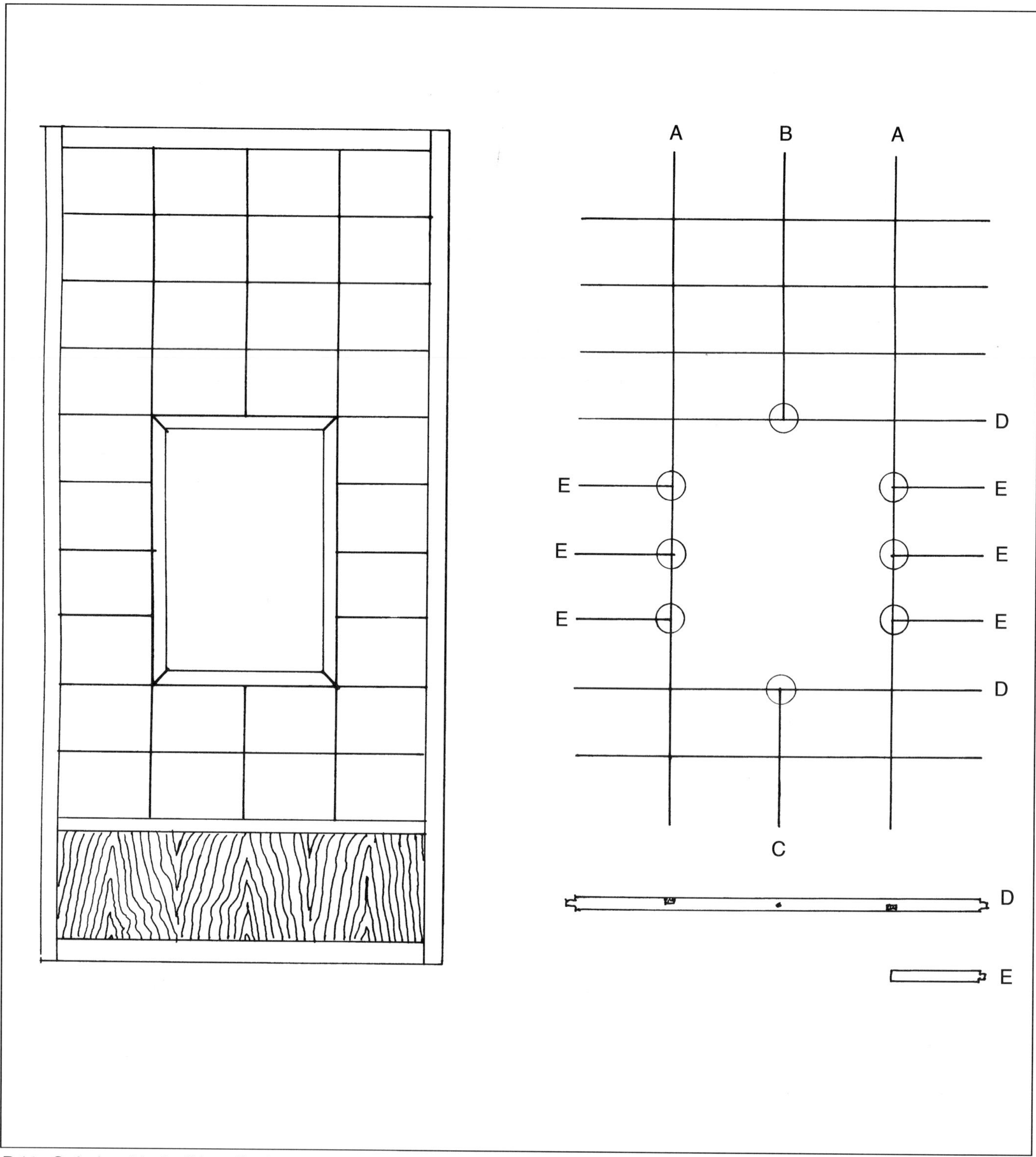

D41 *Gakubuchi-shoji kumiko* layout.

Varieties of Shoji

Now I will introduce you to some types of *shoji* that require quite advanced skills to make. Thus, before you attempt to make these *shoji,* make sure you understand the fundamental functions and master the construction of the common *shoji.*

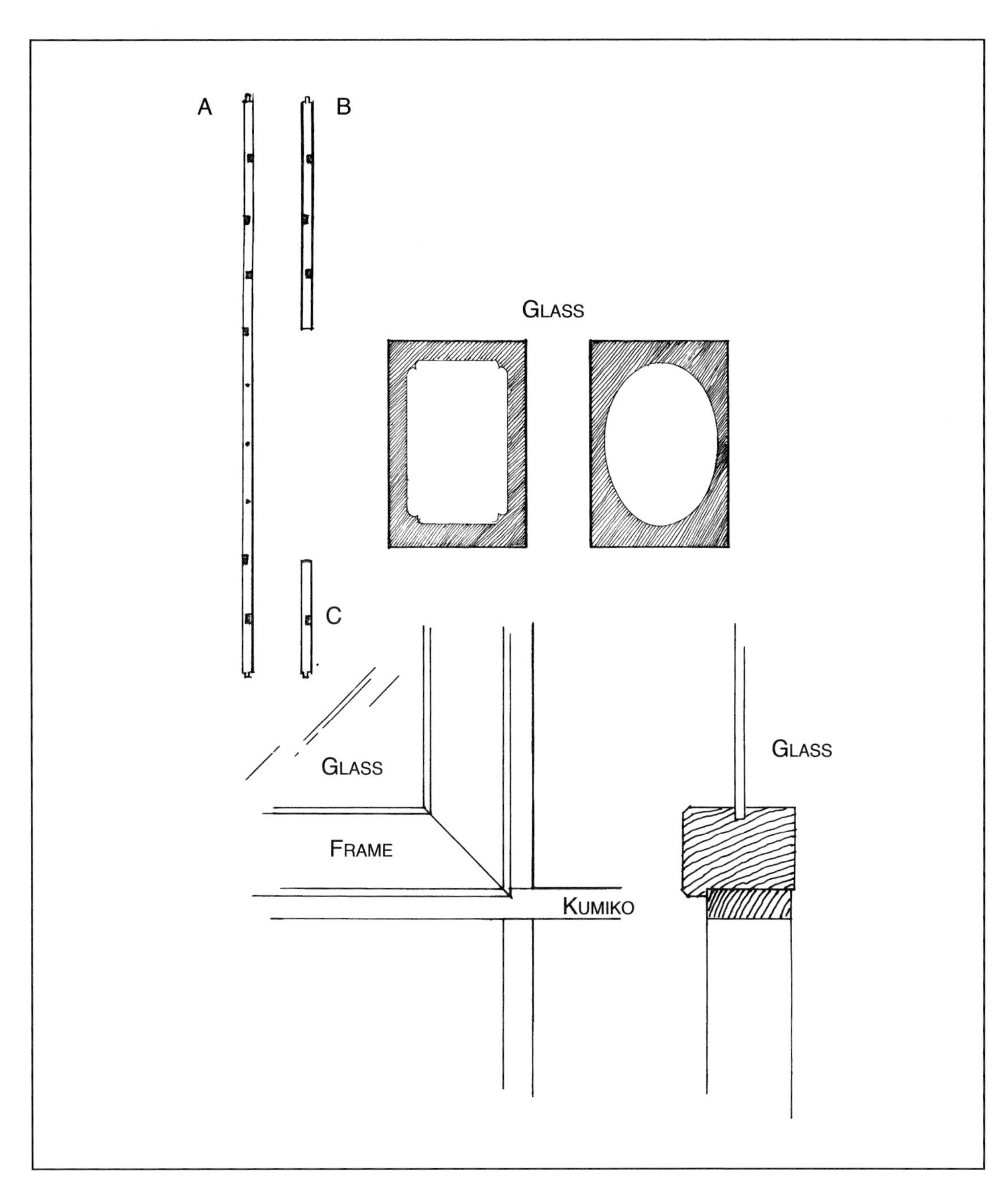

D42 Detail of glass frame.

GAKUBUCHI-SHOJI

PICTURE-FRAME WITH GLASS PANEL

This *shoji* usually faces the exterior of a Japanese house. One can then peek through the framed glass and into the garden. The difference between the *kumiko* cutting of the common *shoji* and this *shoji* is illustrated at the circled points, where B, C, and E are squarely cut (D41). Use small nails from the inside with a dab of glue for attachment. Some craftsmen choose to use blind mortises at these points. If you like you can try, but I would not recommend it. This drawing is showing *mino*-size paper.

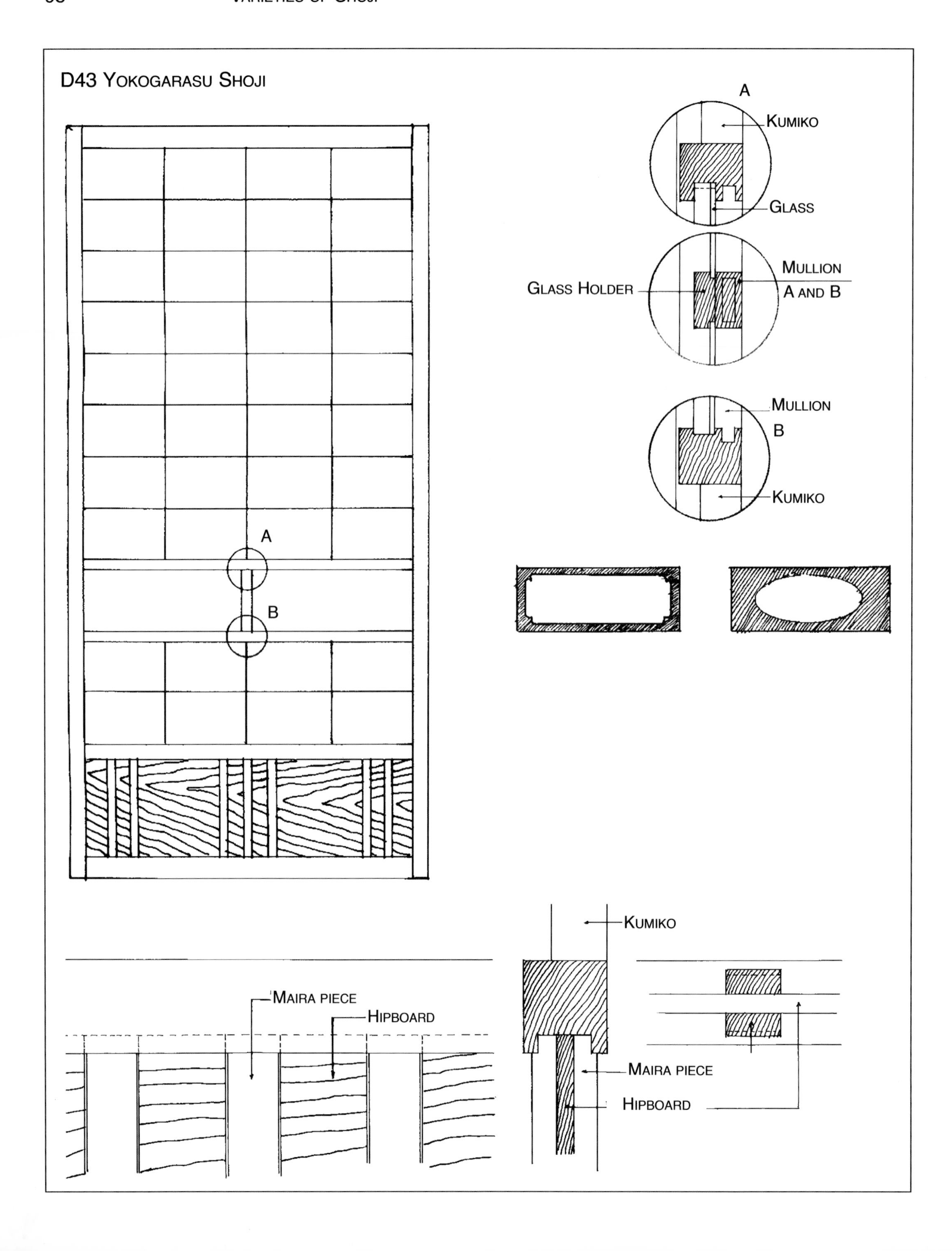
D43 YOKOGARASU SHOJI
A
KUMIKO
GLASS
GLASS HOLDER
MULLION
A AND B
MULLION
B
KUMIKO
A
B
MAIRA PIECE
HIPBOARD
KUMIKO
MAIRA PIECE
HIPBOARD

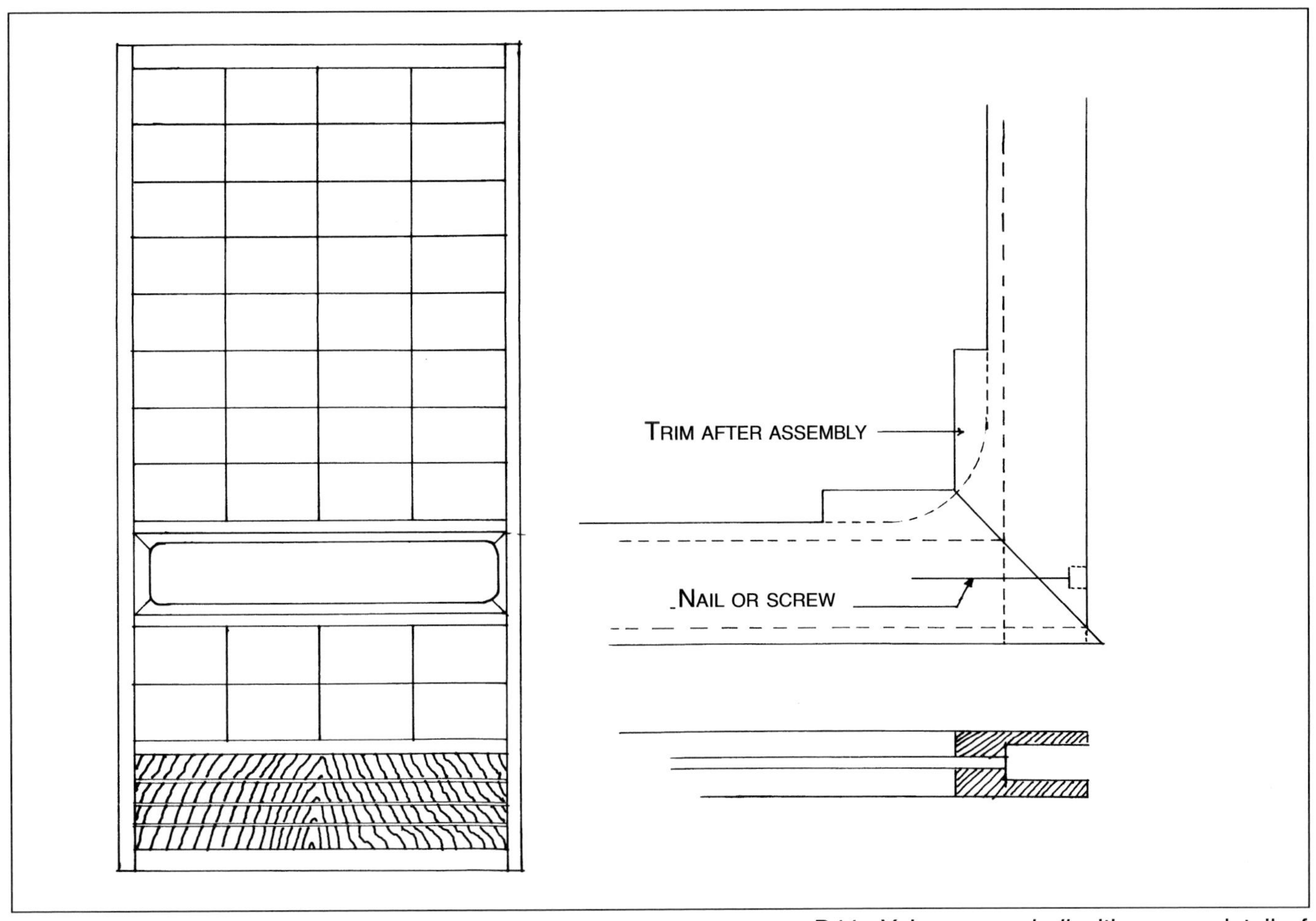

D44 *Yokogarasu-shoji* with corner detail of glass frame.

YOKOGARASU-SHOJI

HORIZONTAL-GLASS *SHOJI*

There are two or three different types of this particular *shoji*. Some, as shown in D43, are used just like the picture-frame *shoji* to face the exterior, but they are also used between the kitchen and other rooms. The other type, as shown in D44, is used only by the entrance door. Inside, by the base of the door, is a concrete floor measuring about 6'x6' or 6'x9'. One then travels up two steps to a *tatami* floor. Right before the *tatami* floor stands this elegant horizontal-glassed *shoji*. The *shoji* shown in D43 usually has one or two pieces of horizontal glass; the *shoji* in D44 has only one horizontal glass.

Let's start with the *shoji* that has one long glass (D44) This is a *hanshi*-sized *shoji*. The difference between the common *shoji* and this one is the two extra middle rails, glass with frame, and hipboard. The framed glass goes into the top and bottom grooves of the middle rails and both sides of the frame fit tightly against the stiles. It is important that once the glass frame is assembled into the grooves, the face side of the frame be all even in width (D45).

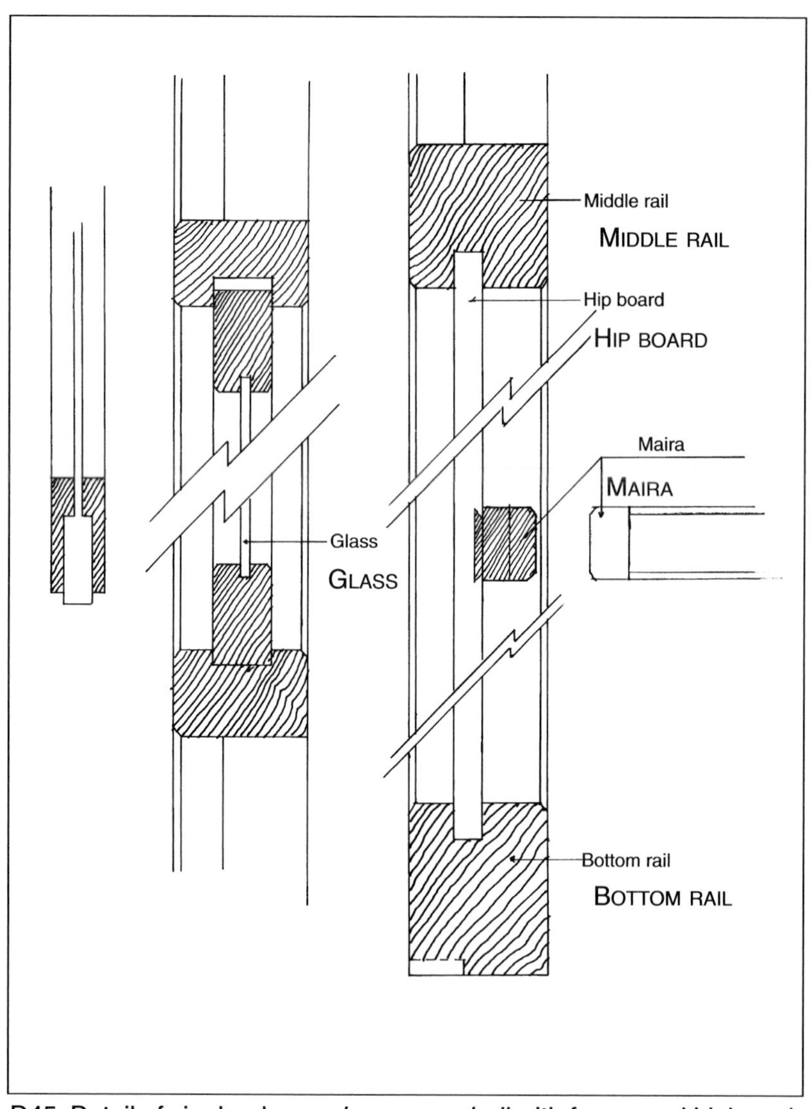

D45 Detail of single-glass *yokogarasu-shoji* with frame and hipboard.

The *shoji* in D43 can have one horizontal glass, just as the *shoji* in D44, or two pieces. If there are two pieces of glass, they are placed directly into the grooves of the middle rails, in between a mullion, being the face, and glass holders facing the back of the *shoji*. Often this glass has frosted landscape designs or a picture frame.

These types of *shoji* have a special hipboard. Both types of the *shoji* shown in D43 have a horizontal-grained hipboard and seven vertical *maira,* thin pieces of wood on the panel with intervals in between the groups of two and three *maira.* These are placed in front and back of the hipboard. The *shoji* in D44 also has *maira,* but the hipboard grain is vertical as in a common *shoji.* It has three horizontal *maira* placed across the grain, with intervals in between them. These *maira* are placed only on the front of the hipboard and are held with a sliding dovetail.

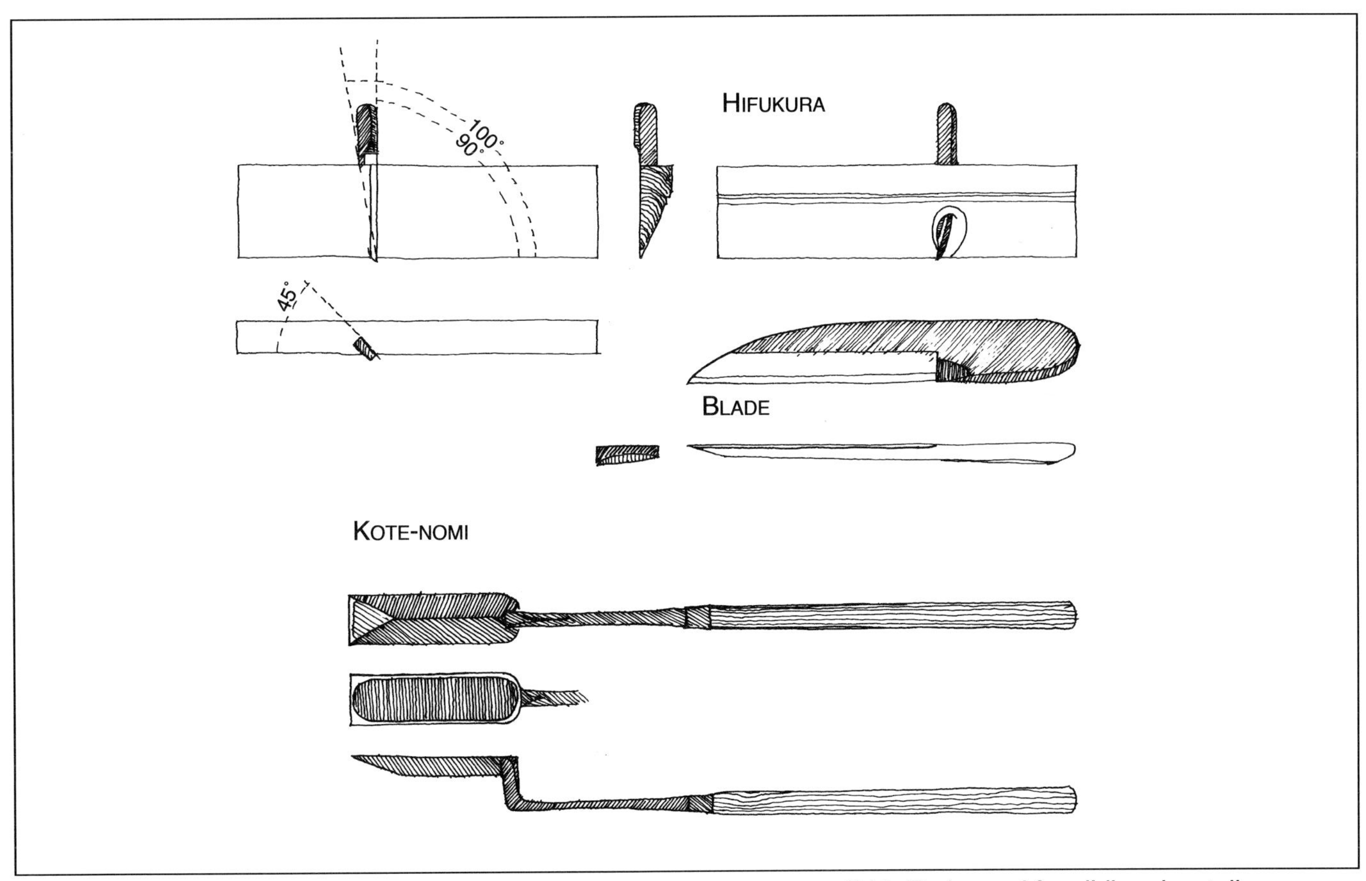

D46 Tools used for sliding dovetail.

SLIDING DOVETAIL

Regular or sliding, the Japanese call the dovetail joint *ari*. Rarely did *tategu-shi* use the common dovetail, but quite often they did make use of the sliding dovetail. It was mainly found on cutting boards, pot covers, table tops, panel doors, and here, on the hipboard. The sliding dovetail was frequently used until the late 1950s. From then on, most objects made of solid wood were replaced by new materials.

Today Japanese craftsmen have two or three types of tools to cut the sliding dovetail joint (D47, page 102). However, when I was young, we did not have these tools, so we made this joint in a different way. Today I own these tools, but nevertheless I am still used to my old ways. In addition, I feel I have more freedom to make any angles I please.

First I mark on the hipboard the width of the *maira*, but a hair narrower. This will be a straight sliding dovetail, not a tapered one, and depending on the thickness of the board, the depth will be about 1/8" to 3/16" (3mm to 5mm). Use a marking gauge to scribe the depth, and do not break the gauge immediately, for you are going to have to use it later on. The angle of the tail is 25 to 27 degrees. Do not make the angle too large, for it will weaken the neck. Likewise, if you make the angle too small,

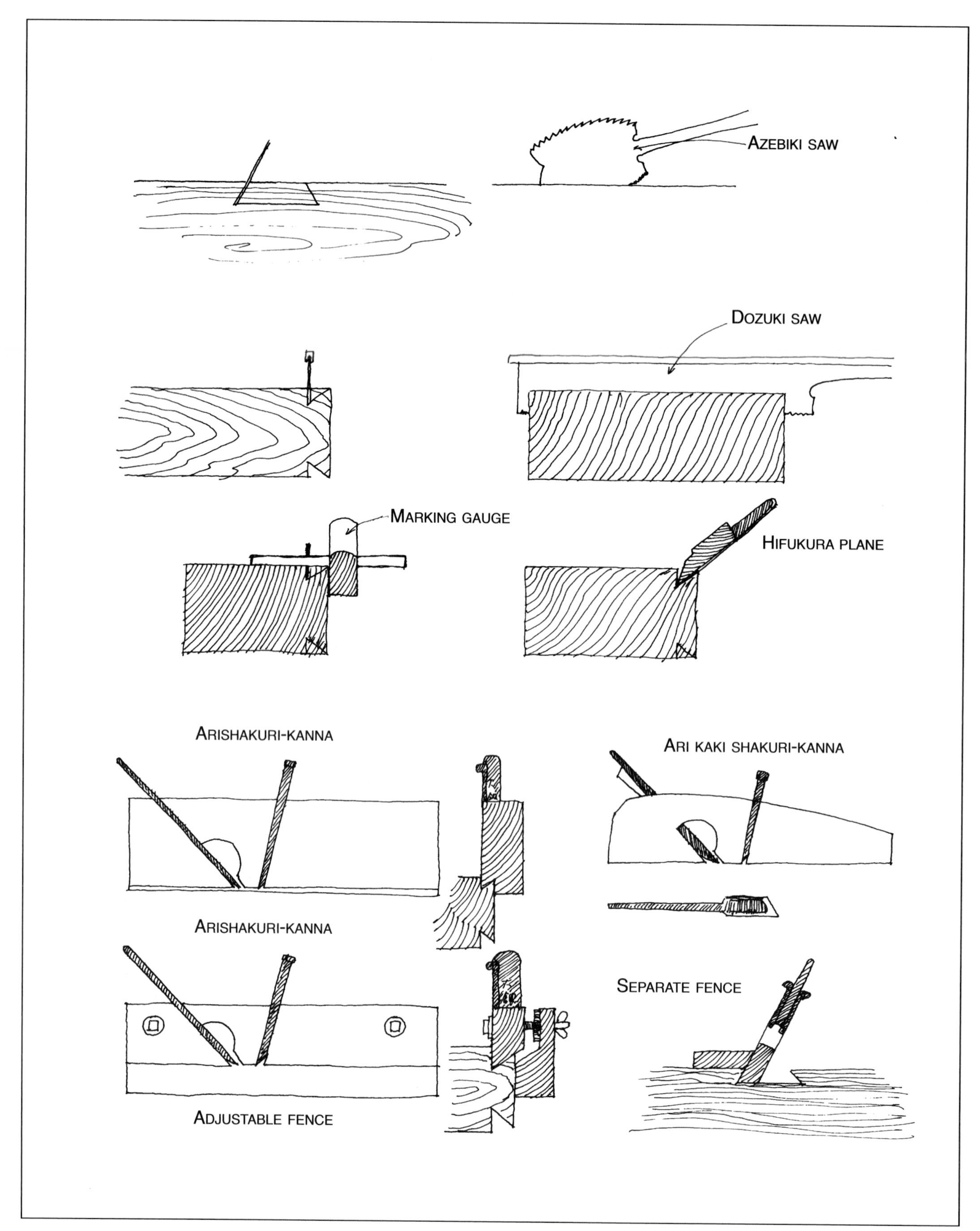

D47 Common way to make a sliding dovetail joint.

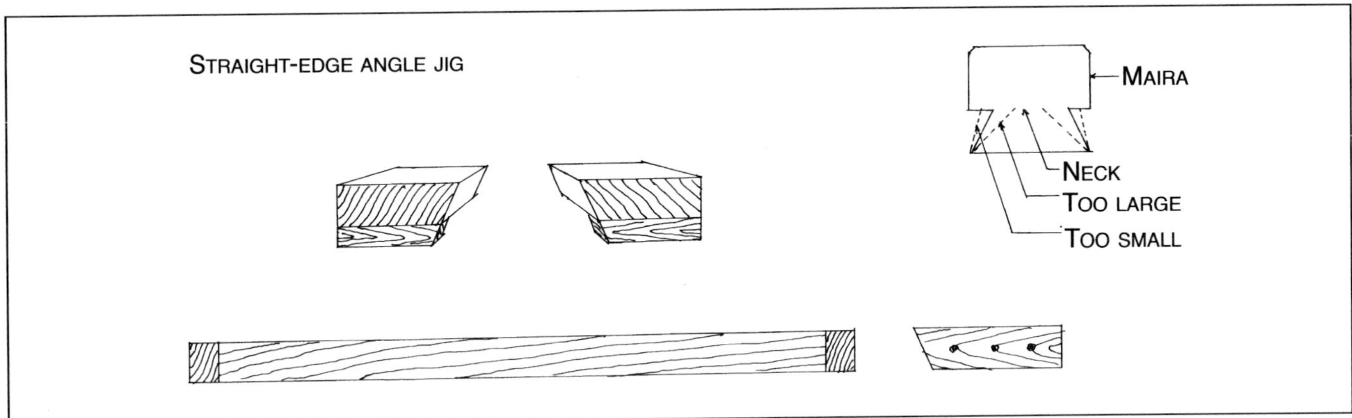

D48 Tools and jigs for sliding dovetail.

because of the *maira's* length, the tail might come out of the groove, despite the mortise and tenon holding both extremities (D48).

We use the *azebiki* saw or *dozuki* saw to cut the cheeks of the groove, and I just follow the angle lines on both edges with my free hand (D47). However, you can use a 2" or 3" (50mm or 75mm) jig or a long straight-edge angle jig held by clamps (D48). This might be a bit easier to start or finish the cut. After finishing the sawed cut I use a mortise chisel, bevel down, to roughly take the waste out. Then I use the *kote-nomi* (trowel chisel), to clean the bottom evenly. This push chisel is especially useful for cleaning up the grooves of sliding dovetails. The top of the blade is shaped like a long slope, so it is easier to cut and clean the sharp corners of the joint. The offset handle allows the user to clean up long, deep grooves without hitting the handle or hand against the wood (D46, page 101). After having gone halfway, turn the wood around and start the procedure over. Check with a ruler to be sure the groove is flat. Make the center a little lower to be certain both mouths clearly touch the ruler.

Then use the same gauge, and mark by pressing hard on the *maira*, the sliding piece. Then use the *hifukura* plane to create the desired angle on the sliding piece. This is the traditional tool for such work in Japan. One presses hard on the *maira*, because it will be easier for the *hifukura* blade to follow the mark. In this case, to cut the cheek of the sliding tail I push the plane (photo 216). Furthermore, in this particular work, I make the groove a hair deeper than the tail so that the shoulder will meet the board tightly.

Now the sliding piece is ready to be part of the hipboard, I first hammer the *maira* 5" to 6" (125mm to 150mm) in, then I press down with my feet, or in your case, with your hand, while continuing hammering through the *maira*. The reason for placing weight onto the sliding tail is to prevent the dovetail neck from tearing off from its upper portion.

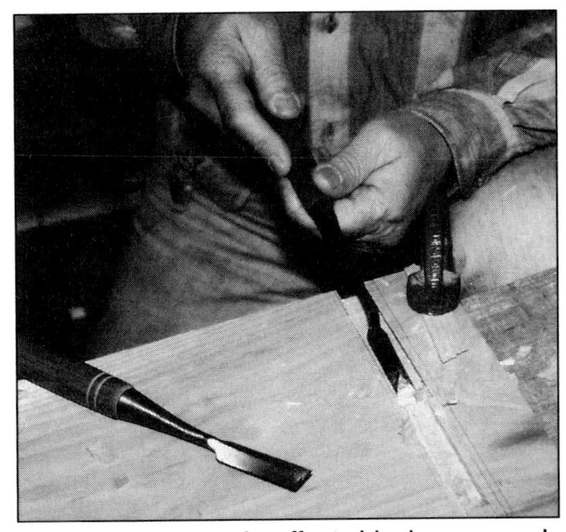

215 The *kote-nomi*—offset chisel—can reach into all parts of the dovetail groove. Check with a ruler to be sure the groove is flat.

216 The *hifukura* plane creates the angle on the sliding piece.

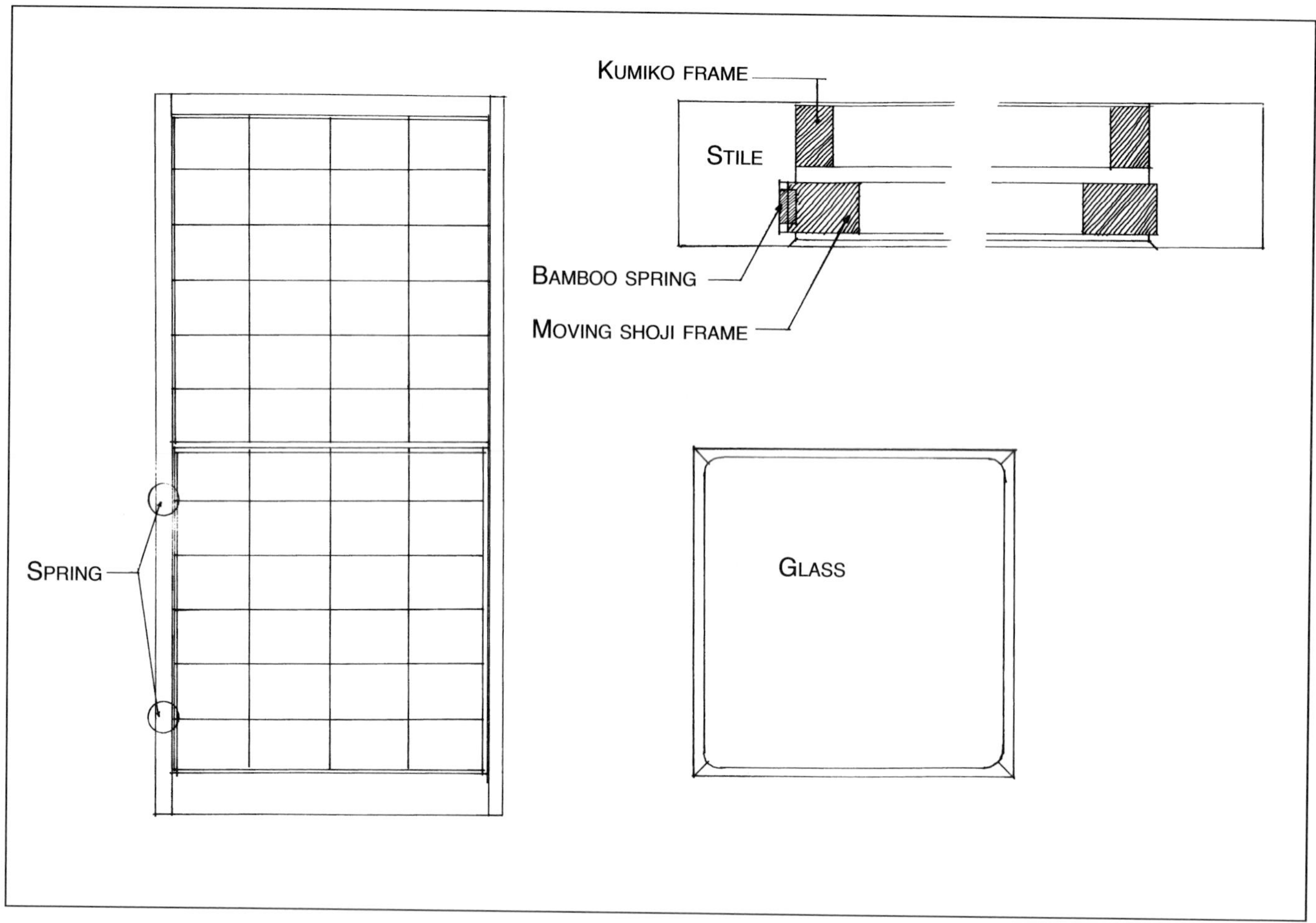

D49 Details of the moving parts of *yukimi-shoji* without hipboard.

YUKIMI-SHOJI—SNOW-WATCHING *SHOJI*

There are two types of this variety of *shoji*. One has a hipboard, the other does not. Since you already know how to use a hipboard, I will explain the type without a hipboard, which we call Western *nekoma shoji* (cat-peeking *shoji*). Western *nekoma shoji* alludes to all *shoji* whose interiors have moving parts, but one also calls this particular *shoji* *yukimi-shoji*. It was developed around Osaka and Kyoto, in the western part of Japan. This type of *shoji* faces the exterior part of the house. Originally when one would open the bottom part of this *shoji*, the exterior air, be it cold or hot, would come through. However, early in this century people started using a fixed glass on the bottom half of this *shoji*. Then, when the bottom part was lifted, one could look outside without feeling a sudden change of temperature.

Before you attempt to construct this *shoji*, I suggest you thoroughly understand the common *shoji* and sharp blades, as well as master the skill to perfectly execute each step of construction. Good material and high skills are most essential. In this *shoji* detail, the upper fixed part has a *kumiko*

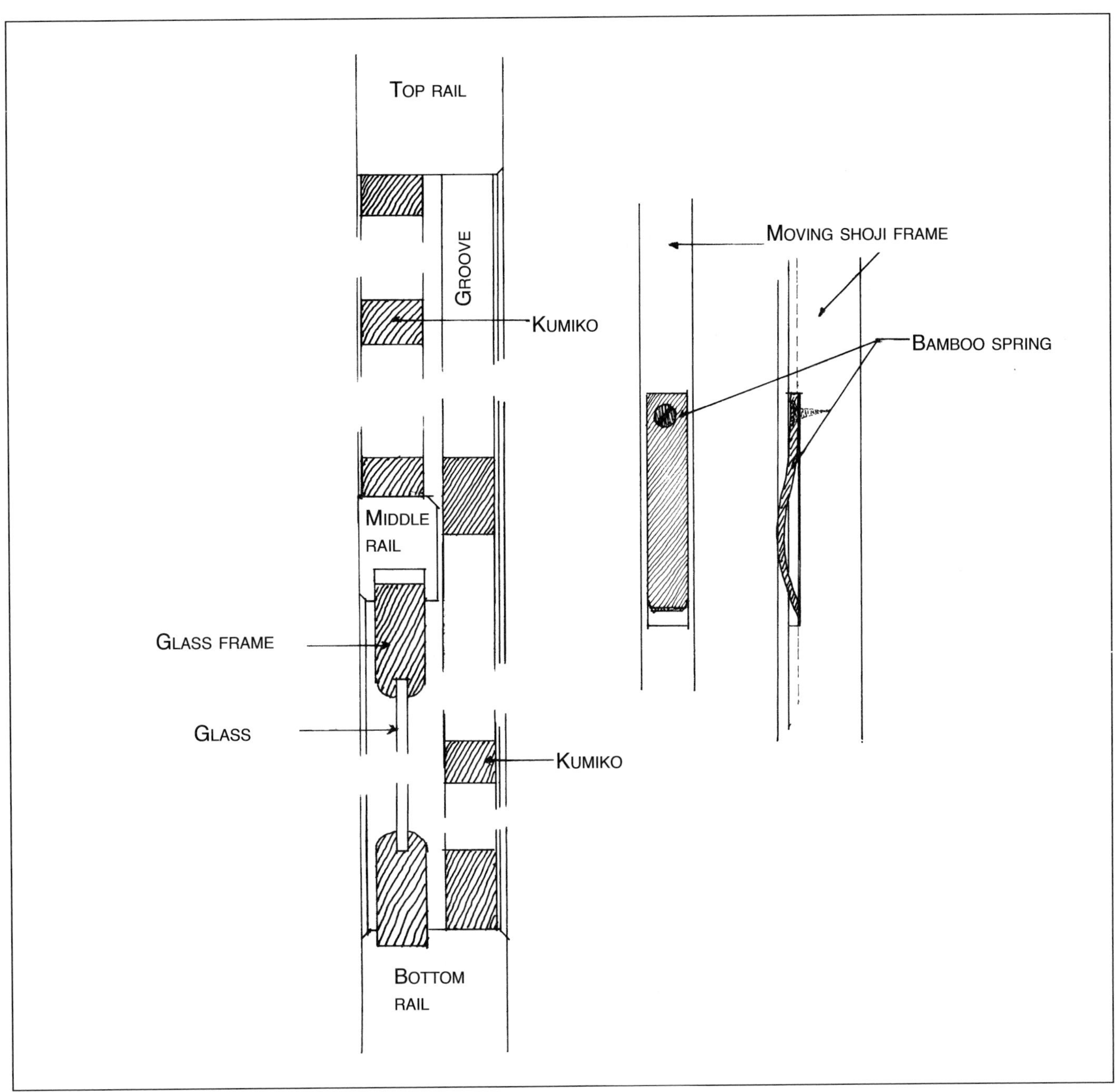

D50 Detail of the moving *shoji* and glass frame installation of *yukimi-shoji*.

frame while the moving part does not. When you lift it up completely, the front *shoji kumiko* and back *shoji kumiko* have to exactly overlap each other. The spring could be made with a clock spring, but traditionally it was made of bamboo. Personally I use bamboo, for it wears out the groove much less than metal does. You can manipulate the bamboo into shape with steam or a candle flame. If you use bamboo, make sure to use the skin—the smooth side. And lastly, the *shoji* stiles and rails all have the same thickness, for they are chamfered on both sides of the *shoji*. Unlike the other common doors, their rails are 1/16" (1.5mm) thinner than the stiles.

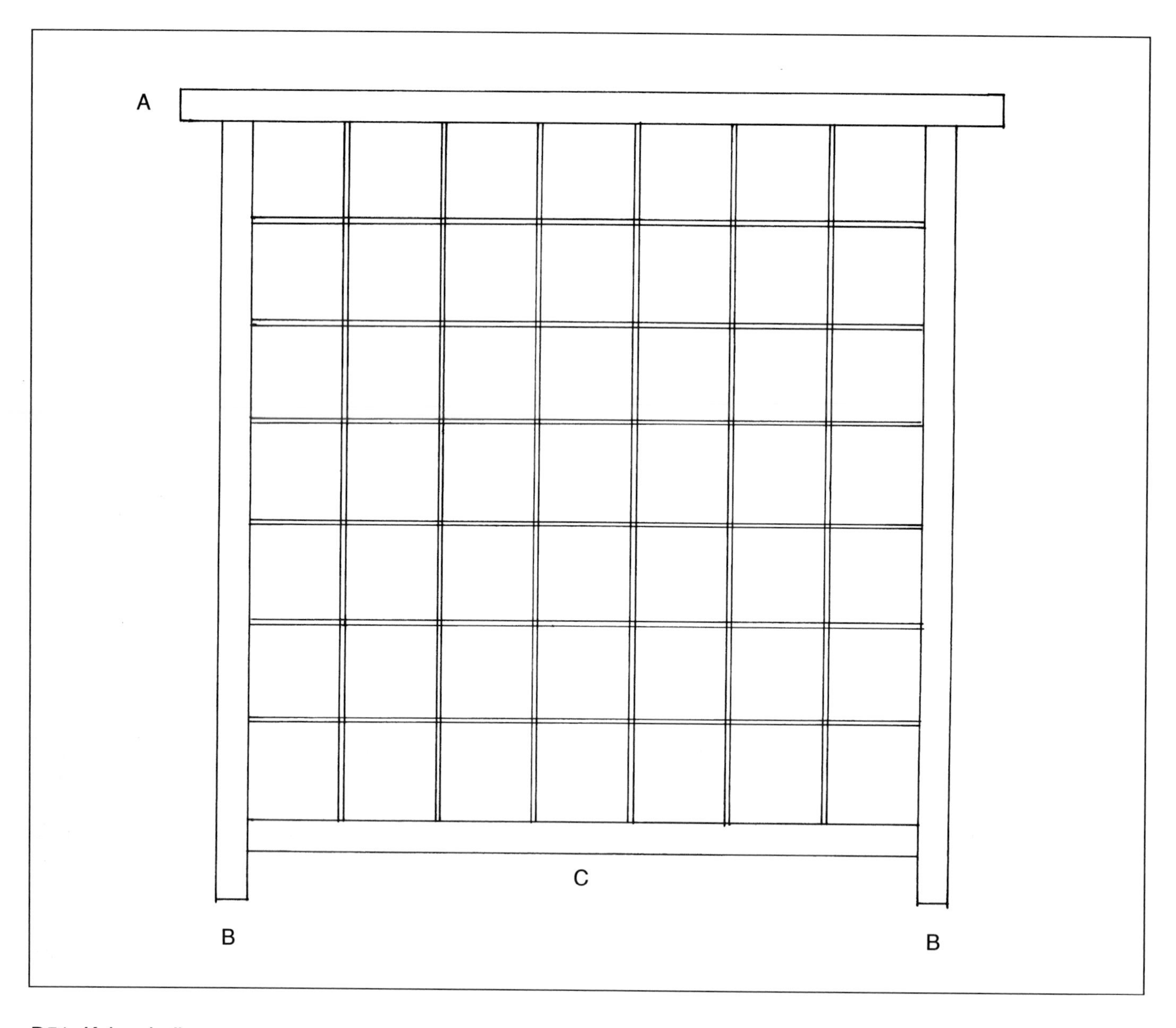

D51 *Kake-shoji.*

KAKE-SHOJI—HANGING SCREEN

Often the Japanese cover windows or glass doors with a screen. Here is a very common way to construct the cover. You can make it in any rectangular size, but most Japanese screens like this are square. If it's square, then mark all materials together (D51). The frame measures 2" x 1" (50mm x 25mm) and the *kumiko* measure 3/8" x 3/4" (9.5mm x 19mm). In this case you can ignore the size of the *shoji* paper, and just cut paper to fit the *kumiko*. Also you do not have to care for the overlapping rule of one *kumiko* size on one frame, for it will most likely not be visible.

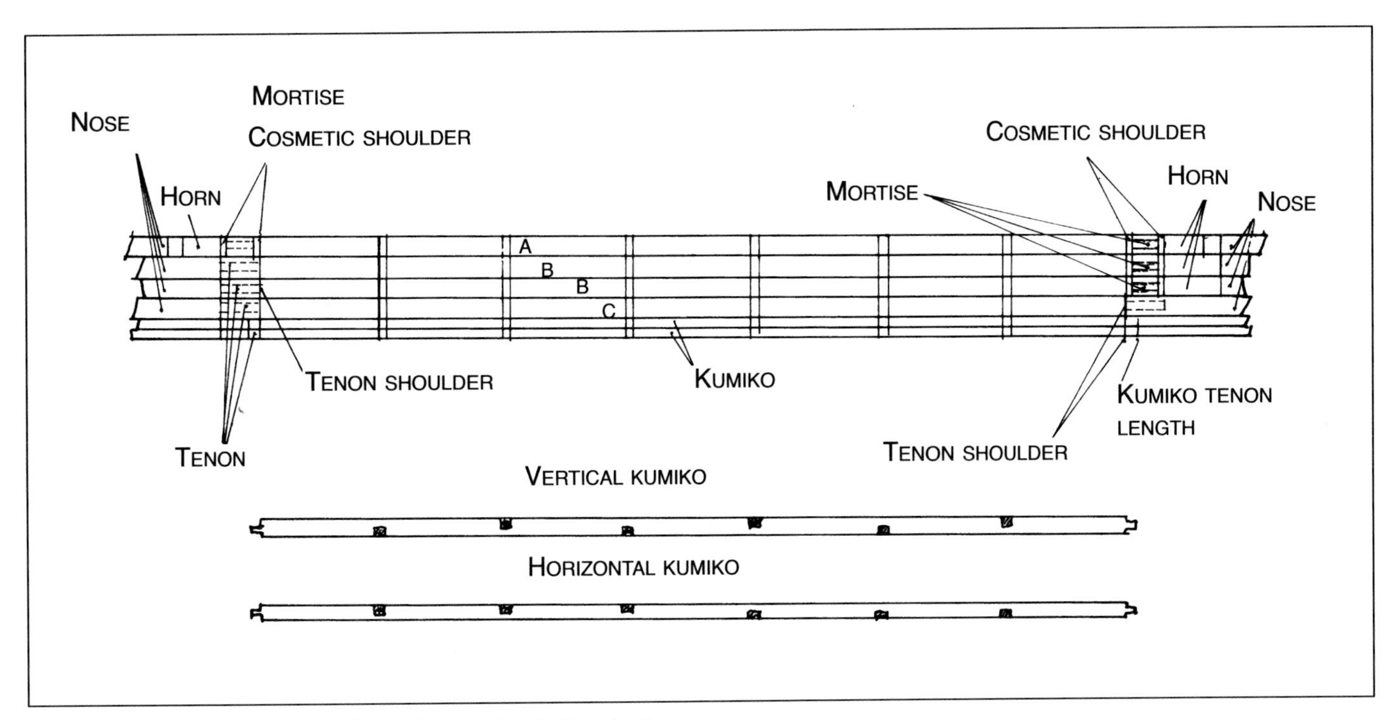

D52 Marking procedure and *kumiko* cut for *kake-shoji*.

217 The hanging screen made of oak. Notice the vertical and horizontal horns.

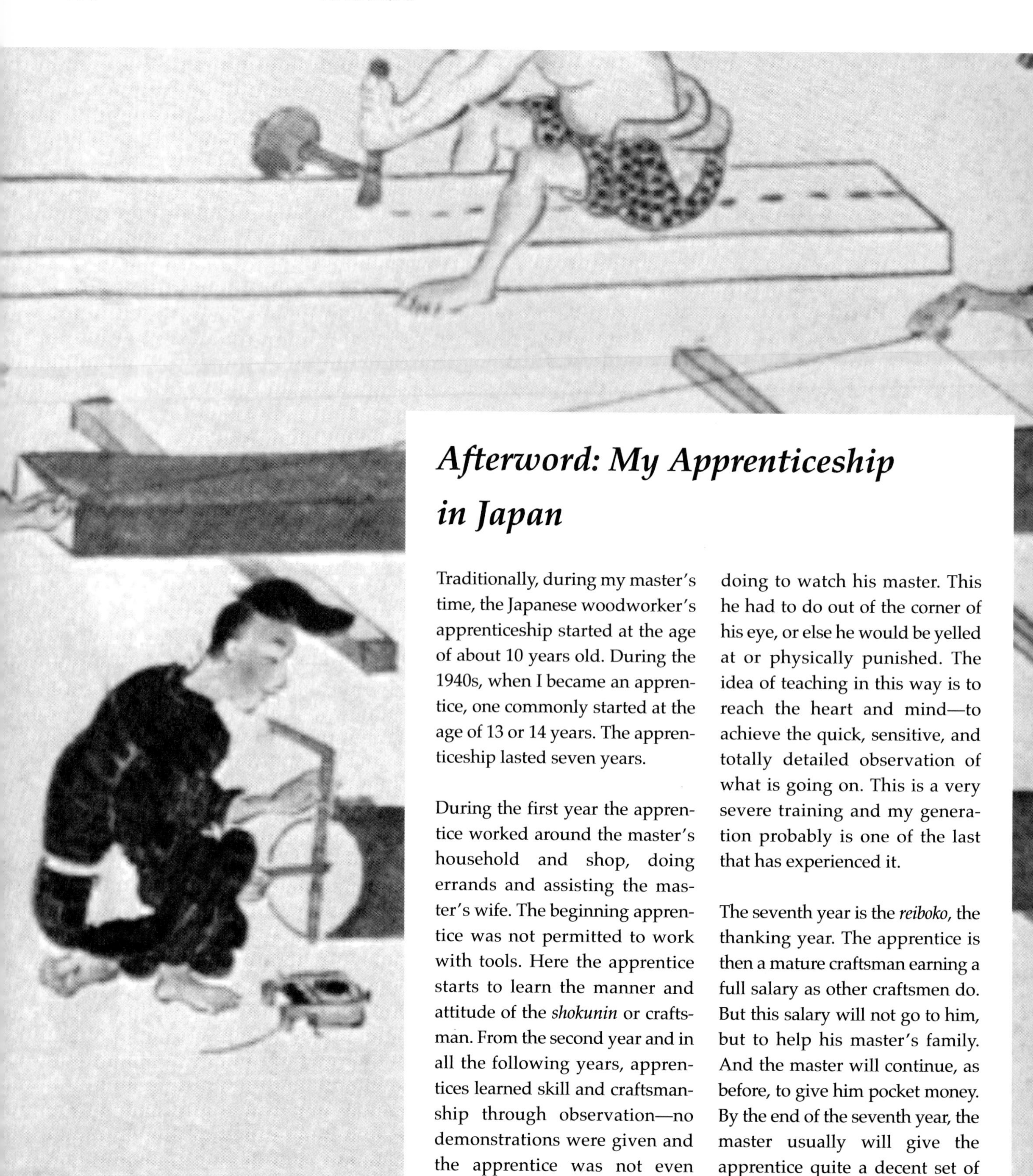

Afterword: My Apprenticeship in Japan

Traditionally, during my master's time, the Japanese woodworker's apprenticeship started at the age of about 10 years old. During the 1940s, when I became an apprentice, one commonly started at the age of 13 or 14 years. The apprenticeship lasted seven years.

During the first year the apprentice worked around the master's household and shop, doing errands and assisting the master's wife. The beginning apprentice was not permitted to work with tools. Here the apprentice starts to learn the manner and attitude of the *shokunin* or craftsman. From the second year and in all the following years, apprentices learned skill and craftsmanship through observation—no demonstrations were given and the apprentice was not even allowed to stop what he was doing to watch his master. This he had to do out of the corner of his eye, or else he would be yelled at or physically punished. The idea of teaching in this way is to reach the heart and mind—to achieve the quick, sensitive, and totally detailed observation of what is going on. This is a very severe training and my generation probably is one of the last that has experienced it.

The seventh year is the *reiboko,* the thanking year. The apprentice is then a mature craftsman earning a full salary as other craftsmen do. But this salary will not go to him, but to help his master's family. And the master will continue, as before, to give him pocket money. By the end of the seventh year, the master usually will give the apprentice quite a decent set of tools and will include one of his

own tools. Sometimes he will even make him a new tool box. Finally a dinner ceremony is prepared and all drink to the end of the apprenticeship. After this the young craftsman may stay with his master, but often he will move out from his master's place and establish his trade elsewhere or travel town to town, province to province, to visit other master craftsmen. There he will stay and work for a short period of time. One calls this path *jungyo* or traveling study—the purpose is to expose himself to new skills, new situations, new tasks, to refine his skills, and to pursue a higher achievement of his craftsmanship. The path reminds me of a similar story, the voyage of Siddhartha. However, this traditional system was all disrupted by the Second World War.

I started my apprenticeship at the age of 16, and my gofer period only lasted six months. For personal reasons I did not have the *reiboko*, also I went at an early stage to other craftsmen's workshops and sometimes worked alone.

After the apprentice's first year he will receive a modest, low-quality set of tools. In my trade the *tategu-shi* makes doors and screens, but my master and I made many other kinds of things like cabinets, tables, and kitchen utensils. By the time the *tategu-shi* apprentice receives his new tools, his first task is to make an *amado* (wooden panel storm door), and if there is no other younger apprentice behind him the gofer job will remain his. The apprentice starts by making the *amado* because it is least visible to peo-

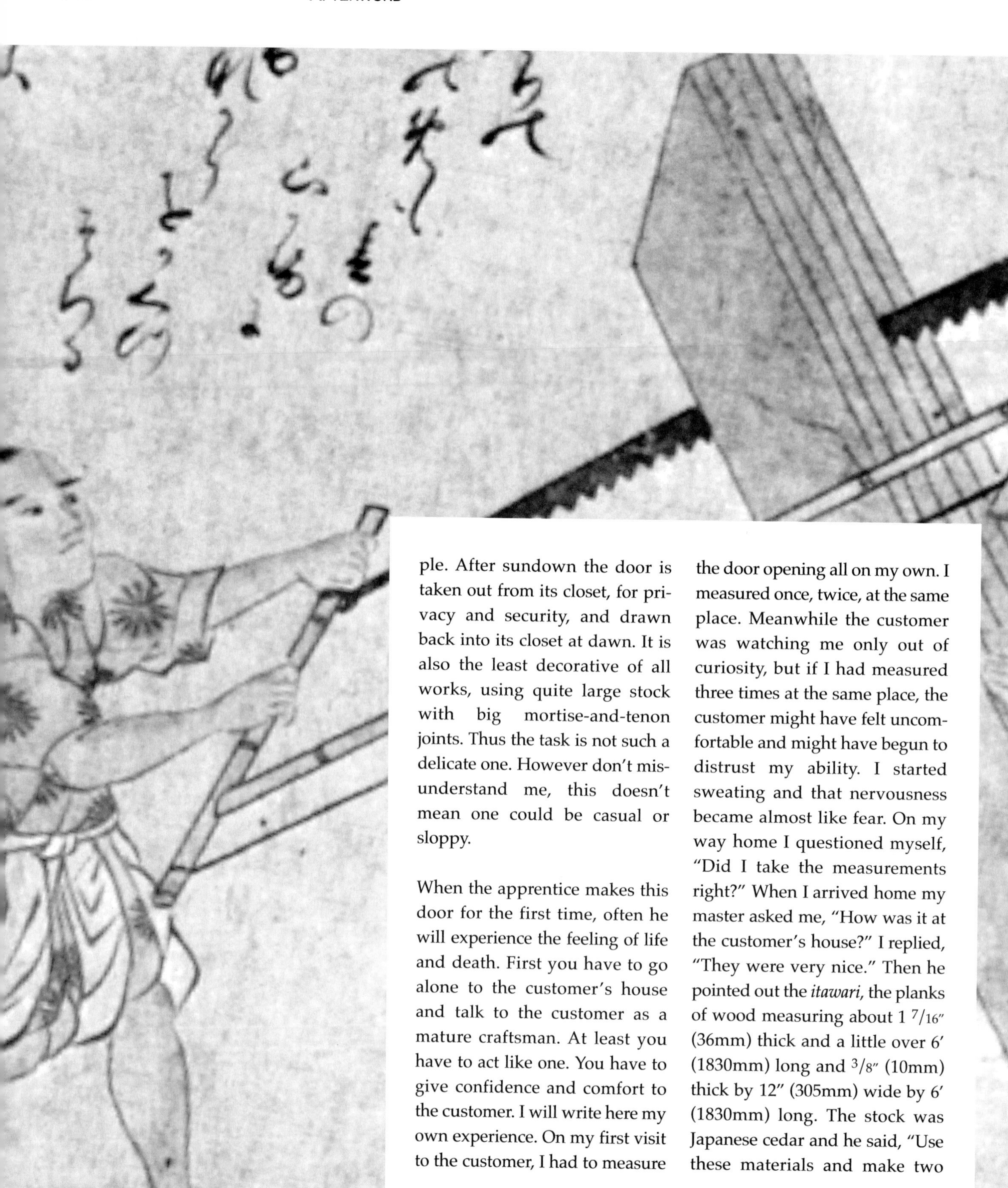

ple. After sundown the door is taken out from its closet, for privacy and security, and drawn back into its closet at dawn. It is also the least decorative of all works, using quite large stock with big mortise-and-tenon joints. Thus the task is not such a delicate one. However don't misunderstand me, this doesn't mean one could be casual or sloppy.

When the apprentice makes this door for the first time, often he will experience the feeling of life and death. First you have to go alone to the customer's house and talk to the customer as a mature craftsman. At least you have to act like one. You have to give confidence and comfort to the customer. I will write here my own experience. On my first visit to the customer, I had to measure the door opening all on my own. I measured once, twice, at the same place. Meanwhile the customer was watching me only out of curiosity, but if I had measured three times at the same place, the customer might have felt uncomfortable and might have begun to distrust my ability. I started sweating and that nervousness became almost like fear. On my way home I questioned myself, "Did I take the measurements right?" When I arrived home my master asked me, "How was it at the customer's house?" I replied, "They were very nice." Then he pointed out the *itawari*, the planks of wood measuring about 1 7/16" (36mm) thick and a little over 6' (1830mm) long and 3/8" (10mm) thick by 12" (305mm) wide by 6' (1830mm) long. The stock was Japanese cedar and he said, "Use these materials and make two

amado." Traditionally the *tategushi* never uses blueprints or text, everything is planned in his head. However, while doing my chores as a gofer, only with the corner of my eyes did I see my master making doors. He never told me or showed me how to make this door, yet this day came, he was expecting me to know everything about the *amado*.

A Japanese craftsman's common saying is, "Technique is not taught, one has to steal it." I did not have any time to waste, I took the *itawari* out, studied the grain, and checked if there were any defects. Then I started to think which part of the plank is good for the stiles, since you should always start by marking the longest pieces first. Then I marked top, bottom, and middle rails. My body and arms were very busy moving but half of my mind didn't know what I was doing. I knew my master was working on his own job, and that he was watching me sharply from the back of his head, he had an extra pair of ears and X-ray eyes. He knew everything I did and could read through my mind. I did not have any idea how wide the rails and stiles should be. I pretended going to the bathroom and on the way I checked the measurements of the stiles and all three rails of our own door.

I finished the layout on the *itawari* and then started ripping the pieces from the plank. A mature craftsman would have to finish one *amado* a day. Now, I had to make two doors for this customer. I was working very hard, but it was not fast enough. Later I became able to make three such

doors in two days. But this day I did not take lunch or tea breaks, and worked even after dinner. At tea time my mother gave me a sweet potato. I could not eat it.

The next day I started very early in the morning for I knew I was behind schedule, but my master didn't say anything, which made me very uncomfortable. I wished that at least he would yell at me and question what I was doing. The day after, at sundown, I gave the final planing and I was ready to assemble the doors. By then my heart was beating faster. As I was assembling the first frame I noticed with horror that both top mortises had cracked despite their horns. I took the tenons out, pared them and the mortise cheeks, and as a result the tenons became a little too loose. But by now I could not do anything about it, the mortises were deeply cracked. And I had to hurry for the glue was drying. I reassembled them, glued the cracked mortises, and clamped them with *hatagane*, Japanese small clamps.

The second frame I handled correctly. The mortises and tenons were made carefully and were not too loose, but during assembly, despite all my efforts, one top rail mortise cracked. When my master came home, the two door frames were standing with many clamps sticking out from the horns. Just once and only from a distance did he look at the frames, I could not see his face, and no sound was heard. I felt his lips closing tight and every one in the house felt his displeasure in the air.

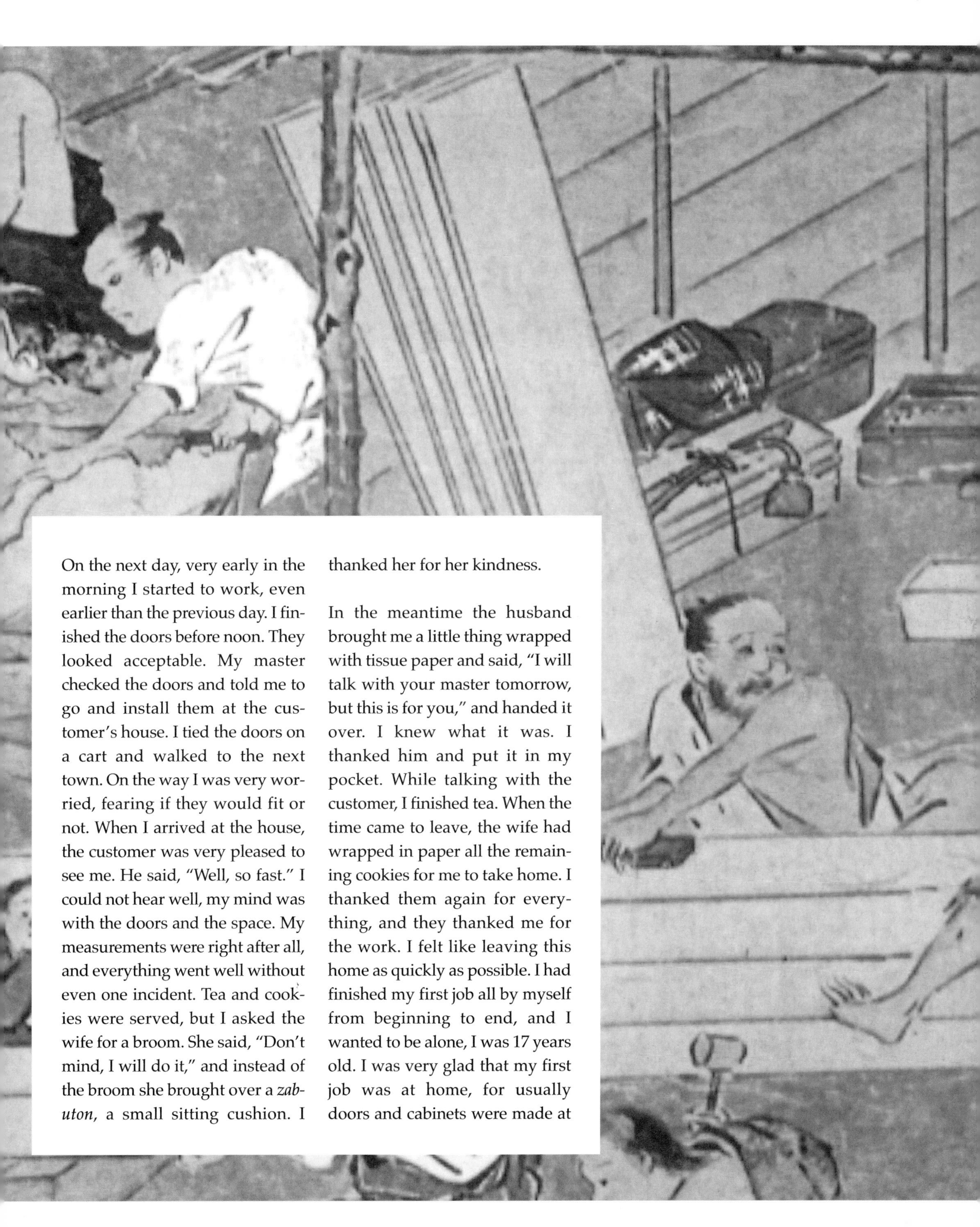

On the next day, very early in the morning I started to work, even earlier than the previous day. I finished the doors before noon. They looked acceptable. My master checked the doors and told me to go and install them at the customer's house. I tied the doors on a cart and walked to the next town. On the way I was very worried, fearing if they would fit or not. When I arrived at the house, the customer was very pleased to see me. He said, "Well, so fast." I could not hear well, my mind was with the doors and the space. My measurements were right after all, and everything went well without even one incident. Tea and cookies were served, but I asked the wife for a broom. She said, "Don't mind, I will do it," and instead of the broom she brought over a *zabuton*, a small sitting cushion. I thanked her for her kindness.

In the meantime the husband brought me a little thing wrapped with tissue paper and said, "I will talk with your master tomorrow, but this is for you," and handed it over. I knew what it was. I thanked him and put it in my pocket. While talking with the customer, I finished tea. When the time came to leave, the wife had wrapped in paper all the remaining cookies for me to take home. I thanked them again for everything, and they thanked me for the work. I felt like leaving this home as quickly as possible. I had finished my first job all by myself from beginning to end, and I wanted to be alone, I was 17 years old. I was very glad that my first job was at home, for usually doors and cabinets were made at

the customer's house, using their material. I am sure that it was my master's calculated plan.

The *ma-shoji* job came to me a little over a year later. Usually after making the *amado* the apprentice's next project is to build the *ma-shoji* as explained in this book. This time I was alone in the customer's house making the *ma-shoji*, and it wasn't any easier than the first time, because there were many different kinds of pressures and fears. This way craftsmanship is slowly achieved. The craftsmanship is not only a great skill and technique but also a right attitude. You can learn technique by reading or looking at drawings, but the main part of the craftsmanship is made of concern and consideration. Good technique or beautiful work is just the result of great craftsmanship. To me a craftsman without craftsmanship is like a beautiful-looking bread without nutrients.

How to achieve this craftsmanship without the beatings from a master? Well, you have to begin by having a right attitude, the attitude of a craftsman's social obligation and responsibility, and you should also have a self-discipline of eyes and hands. These are the essential elements for achieving craftsmanship. When you have an opportunity to watch an accomplished craftsman at work, watch and absorb how he or she works, not how they make things.

I wrote this book as part of an introduction to Japanese tradition. I strongly emphasized the

manual aspect of the process, although obviously you can make the objects shown with very simple small power tools. However, deep in my heart I wish you will absorb the craftsman's care and consideration, then, even if you're using power tools, I'm sure your attitude will reach the level I'm talking about. Here I will retell a story I wrote in my previous book.

Chiyozuru Korehide, a great master blacksmith of this century, had two apprentices. Both also became great blacksmiths of plane blades. Although Chiyozuru-Sadahide and Chiyozuru-Nobukuni had the same master, they developed philosophies and styles quite different from one another. Chiyozuru-Sadahide kept to the traditions of the blacksmith, from his tools to the hand-hammering method he employed. Chiyozuru-Nobukuni, on the other hand, broke with tradition to use machinery in his work. When Yoshio Akioka was researching his book on hand tools, "*Nippon No Tedogu*," he made a visit to Chiyozuru-Nobukuni. He commented that although he was a bit disappointed to discover that Chiyozuru-Nobukuni used almost no hand tools, he was nevertheless enlightened by the master's conversation, sprinkled with words seldom used by traditional blacksmiths—words such as "pyrometer," "micron," and "flow-line data." Akioka reports that when Chiyozuru-Nobukuni discussed good finishing stones and the hours of sharpening necessary, he did so in the context of making "a

cutting edge down to about five microns." Further, Chiyozuru-Nobukuni asserted proudly that although scientists and scholars recommended the hardness of the cutting edge be stopped at RC64 (a hardness of 64 on the Rockwell C scale) or else risk crumbling, his was RC67 without crumbling. Today many manufacturers use machinery to produce tools, but they do so only because the process is faster and much cheaper; the result is a product that does not demonstrate quality. Chiyozuru-Nobukuni, on the other hand, understood the possibilities of machine production for better quality at higher speed and a greater degree of accuracy. For instance, he applauded and took advantage of the strength and accuracy of machine-hammering thick steel. Akioka's conclusion is that what Chiyozuru-Nobukuni learned from his master was enhanced by the knowledge of scientists and scholars. As a result, Chiyozuru-Nobukuni's work transcended both the traditional and the modern and became unique in itself.

I hope that this book will enrich your understanding of Japanese woodworking as well as the craftsman's attitude and will be the beginning of a delightful adventure.

Toshio Odate
Woodbury, CT
March 1999

Index